NORMANDY BATTLES

D-DAY

In the early morning darkness of 6 June 1944, the largest armada of ships the world has ever seen heaves to off the Normandy coastline. Aboard, thousands of Allied soldiers wait in readiness for their date with destiny. Months of planning, training and preparation are now behind them. On this day they will undertake the greatest amphibious assault in history, and on their success or failure hangs the very fate of the liberation of Europe from the jackboot of Nazi domination. D-Day is finally here!

Operation Overlord

In November 1943, following months of negotiations, the British and American Governments finally agreed to a full-scale invasion of German-occupied France—Operation Overlord—provisionally scheduled for May 1944. In December, US General Dwight D Eisenhower was appointed commander of Supreme Headquarters Allied Expeditionary Force (SHAEF) charged with planning the invasion. Under his command General Sir Bernard Montgomery's 21 Army Group consisted of the US First Army under General Omar Bradley and the British Second Army under Lieutenant-General Sir Miles Dempsey. The vast forces involved meant that not all could be landed at once, so spearheads would have to land on the invasion beaches and push inland clearing the way for others to follow.

Target Normandy

The Allies chose Normandy for the landings, rather than the shortest route across the English Channel from Dover to the Pas de Calais. Hitler himself suspected that Normandy would be the site of any invasion but, unusually, allowed himself to be persuaded otherwise by his generals. To reinforce this conviction, the Allies launched a major deception plan, Operation Fortitude, using double agents, fake signals transmissions, news stories, broadcasts and dummy encampments. The deception centred on creating the illusion of a First US Army Group, FUSAG, supposedly comprising 30 divisions stationed in south-east England under the

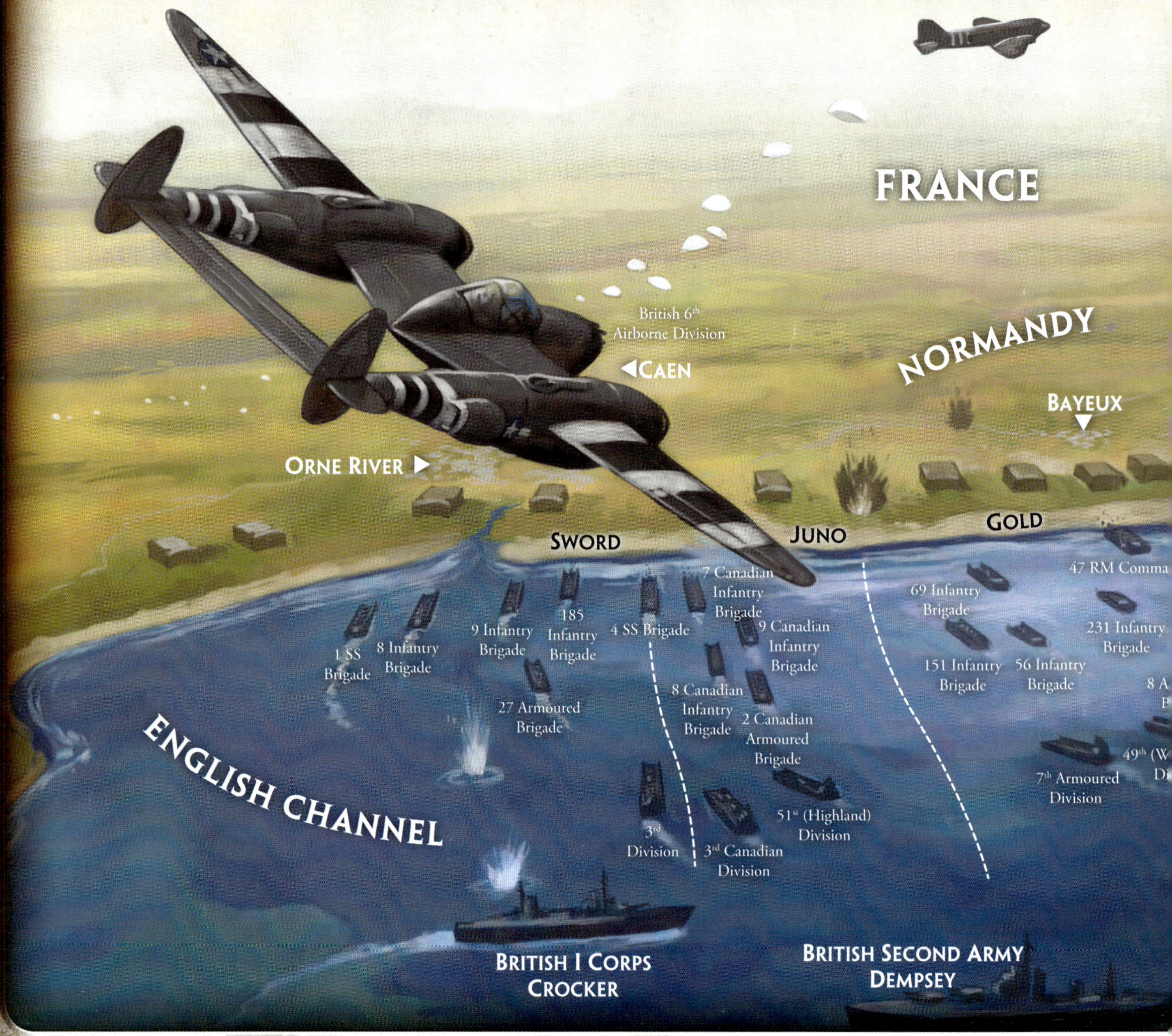

command of General George S Patton. The Germans were completely taken in. Even after the Normandy landings had taken place, Hitler refused to allow reinforcements to be transferred from the Pas de Calais region, believing that the landings were merely a diversionary attack.

The Atlantic Wall

While the Allies laid their plans and marshalled their forces the defenders of Hitler's *Atlantikwall*, the coastal fortifications of North Western Europe, were not idle. Since 1942 *Generalfeldmarschall* Gerd von Rundstedt had been *Oberbefelshaber (OB) West*, commanding all German Forces in France, Holland and Belgium. This included Army Group B, which controlled Seventh Army, defending Brittany and Normandy, and Fifteenth Army in the Pas de Calais region.

In November 1943, command of Army Group B was given to *Generalfeldmarschall* Rommel, the famed 'Desert Fox', with orders to ready the neglected coastal defences for the long expected invasion. Rommel added strong points and took steps to deny landing fields to airborne invaders. Well sited anti-tank obstacles and extensive minefields were constructed to hinder the invaders. However, weaknesses remained. The defences along this part of the coast mostly had to rely on obsolescent ex-French, Russian and Czech weapons, and there was a notable lack of depth in defence once the initial coastal 'crust' was broken.

The Defensive Plan

The German defensive plan involved infantry formations holding the defensive line along the coast, with an armoured reserve held further inland. The bulk of the infantry forces manning the fixed defences were static divisions, comprising two infantry regiments of variable quality supported by two attached *Ost* battalions made up of former Soviet soldiers. The all-important armoured divisions, critical to the success of any counterattack in the event of an invasion, were nominally part of *Panzergruppe West*, directly controlled by *OB West*. However, Rommel did manage to get three armoured divisions placed under his direct control.

This confusing German command structure, and the need to obtain the authority of Hitler himself to move key formations, was to significantly hamper the German ability to react swiftly when the time came. On the day of the invasion, von Runstedt's efforts to move *I SS-Panzerkorps* (I SS Panzer Corps) closer to the invasion beaches had to await Hitler's approval. This was not given until 1600 hours. Even then, Allied air attacks significantly delayed the movement of most reserve formations. In the months preceding D-Day, the Allied air forces had smashed the French railway system, reducing its capacity to move troops to the front. The Germans were forced to commit the few remaining *Luftwaffe* aircraft to its defence against overwhelming odds. The *Luftwaffe* was crushed in the process. On the day of the invasion the Allied air forces would have the skies to themselves.

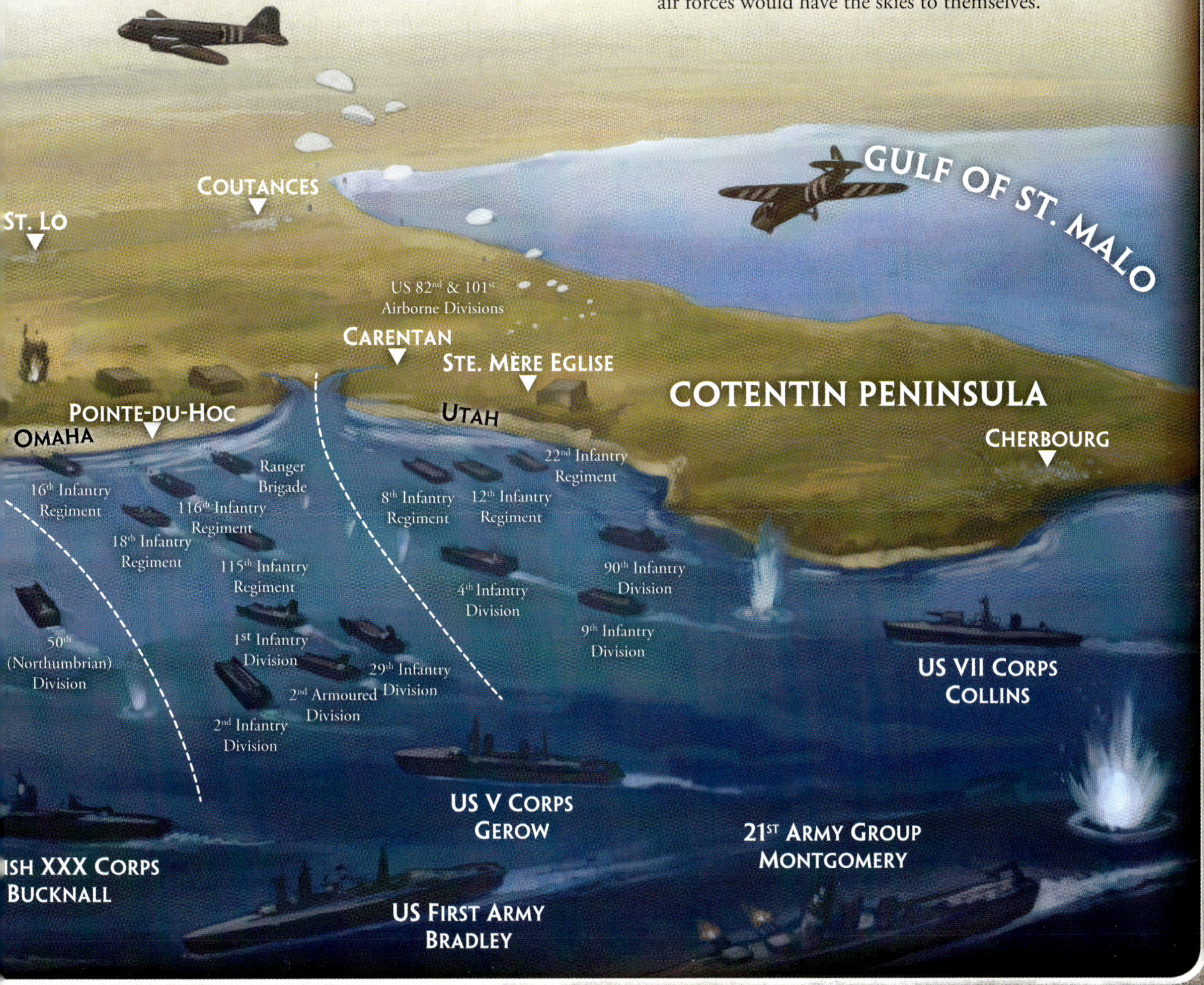

US Army soldiers storm ashore at Omaha Beach, 6 June 1944

D-Day

As D-Day approached, the weather in the English Channel worsened forcing General Eisenhower to postpone the landings by 24 hours. Finally, after consulting the meteorologists Eisenhower made the fateful decision. The weather was not perfect, but it would have to do—the invasion was on.

The poor weather had lulled the defenders into a false sense of security. As the invasion got under way many of the senior German commanders were absent from their posts, attending wargames in Rennes or on leave in Brussels and elsewhere.

Airborne Invasion

In the early hours of D-Day, paratroopers of three airborne divisions—the US 82nd 'All American' and 101st 'Screaming Eagles' Airborne Divisions, and the 'Red Devils' of the British 6th Airborne Division—dropped into Normandy to secure the flanks of the seaborne landings. The more fortunate landed near their drop zones, but many were dispersed as a result of low clouds and anti-aircraft fire.

In the Cotentin Peninsula, on the western flank of the invasion beaches, the US airborne divisions secured key areas inland of Utah Beach. On the eastern flank, the British paratroopers struck at targets between the Orne and Dives rivers. Vital bridges over the Orne and Caen Canal were seized by an audacious glider assault at the outset of the operation.

Although not always successful, the parachute and glider landings proved crucial in confusing and delaying the German defenders, securing inland routes from the invasion beaches and capturing key bridges and crossroads.

At 0300 hours, nearly two thousand Allied medium and heavy bombers hammered the German coastal defences. This airborne onslaught was followed by a massive naval bombardment from seven battleships, 18 cruisers, 43 destroyers, plus gunboats and monitors. A follow-up raid by another thousand American bombers wrought yet more destruction.

Seaborne Landings

Under cover of darkness thousands of landing craft approached the Normandy coastline. The Allied amphibious force would come ashore at five beaches, running from west to east they were codenamed: Utah and Omaha—the landing beaches of the US First Army—and Gold, Juno and Sword—the landing beaches for the British and Canadian troops of the British Second Army.

Utah

Utah Beach, at the base of the Cotentin Peninsula, was wide and flat, and behind the beach was a marshy plain that had been deliberately flooded by the defenders. The Americans elected to land an hour earlier than the British, using the lower tide to overcome the problems of submerged beach obstacles designed to destroy landing craft. At 0630 hours, under cover of a bombardment from rocket-firing landing craft, 8th Regimental Combat Team (RCT) of 4th 'Ivy' Infantry Division led the beach assault. A navigation error put the troops ashore two thousand yards south of the projected landing site. Fortuitously, however, the German defences were even weaker in this sector of the beach.

Supported by amphibious Sherman DD tanks (28 of the 32 launched made it ashore) the infantry quickly overwhelmed *919. Grenadierregiment* of *709. Infanteriedivision*. The 4th Infantry Division secured its objectives at a cost of 200 casualties—far fewer than anticipated. As the rest of the US VII Corps poured ashore, the division linked up with paratroopers of the 101st Airborne Division who had seized the exits from the flooded plain further inland.

Omaha

In contrast with Utah, the going at Omaha Beach was much tougher for the assaulting American troops. Bad weather meant that tides were running higher, swamping landing craft and pushing them onto submerged obstacles. Preparatory fire had missed most of the beach defences, sited on a high bluff overlooking the beach and losses to enemy fire were heavy, with most of the combat engineers and supporting Sherman DD tanks lost before they reached the shoreline.

To further complicate matters, the assaulting troops of 16th RCT, from the veteran 1st 'Big Red One' Infantry Division, and 116th RCT, from the inexperienced 29th 'Blue and Grey' Infantry Division, found themselves facing not only the anticipated *726. Grenadierregiment* of *716. Infanteriedivision*, but also the veteran *914.* and *916. Grenadierregimenter* of *352. Infanteriedivision*, who had occupied the beach defences undetected by Allied intelligence. The assault forces were pinned down on the beach until mid-afternoon, suffering heavy casualties. By nightfall they had advanced no more than 2,000 yards inland.

A few miles to the west, near the Vire River estuary dividing Omaha and Utah, the 2nd Ranger Battalion carried out a daring assault from the sea directly up the cliffs at Pointe du Hoc. The mission was intended to knock out a German coastal battery that threatened the invasion beaches. However, after a successful but costly assault, the rangers discovered that the coastal guns had already been removed by the Germans.

Gold

At 0725 hours the first troops from the British Second Army began landing. At Gold Beach, 69 and 231 Brigade Groups of 50th (Northumberland) Division lead the assault with support from commandos, artillery and specialist armour—the mine-clearing, flame-throwing and engineering 'funnies' allocated to the British beaches. The invaders made good progress against the defenders from *736. Grenadierregiment* of *716. Infanteriedivision*. By early afternoon, all of 50th Division was ashore, with elements of 7th Armoured Division landing behind them later in the day.

Juno

Immediately to the east of 50th Division, it was the task of the 7 and 8 Brigade Groups of 3rd Canadian Division to storm ashore at Juno Beach. The Canadians were supported by the commandos of 4 Special Service Brigade. Mindful of the debacle at Dieppe in 1942 which had cost so many Canadian lives, the Canadians anticipated heavy casualties. In the event, their objectives were taken with comparatively light casualties after hard fighting against elements of *736. Grenadierregiment* of *716. Infanteriedivision*.

By mid afternoon the entire 3rd Canadian Division was ashore, quickly linking up with 50th Division.

Sword

At the easternmost beach of the invasion, Sword Beach, 8 Brigade Group of British 3rd Division led the assault, supported by the commandos of 1 Special Service Brigade. The landings suffered from high tides caused by the bad weather and also tough resistance from German troops of the *736. Grenadierregiment* of *716. Infanteriedivision*. The British troops fought their way through the beach defences and began to exploit inland. On the German side, *716. Infanteriedivision* was practically obliterated, reduced to an effective strength of only two battalions.

3rd Division had been allocated very ambitious objectives, including the capture of the city of Caen, a crucial road and rail junction some ten miles inland. The division duly cleared the invasion beach and linked up with the paratroopers of the 6th Airborne Division, having advanced some 6 miles inland—one of the furthest advances on D-Day.

The unexpected arrival of *21. Panzerdivision* stopped 3rd Division's advance and threw them on the defensive. A strong defence by the British and Canadian divisions prevented *21. Panzerdivision* from exploiting its success, despite reaching the sea between Juno and Sword Beaches. However, its attack had frustrated the initial drive on Caen. Their failure to capture Caen was to have far reaching consequences for the Allies.

Nightfall, 6 June

By nightfall on 6 June the Allies were ashore, but in some places their beachhead was no deeper than 2000 yards. Certain vital D-Day objectives—most notably Caen—had not been captured. Yet enough men and material had been brought ashore that the local German forces could not hope to push them back into the sea. Still, the task that lay before the Allied forces was considerable. They must link up their beachheads, capture Cherbourg (the only major port in the region) to guarantee resupply, and push inland to Caen and St. Lô, before breaking through the difficult *bocage* countryside of Normandy and into the more open terrain beyond.

Canadian troops take cover behind a Sherman tank as they move inland.

THE NORMANDY CAMPAIGN

THE BEACHHEAD

Dawn of 7 June, 1944, found the Allies ashore, but their toe-hold on the Normandy coast was precarious. The German defenders continued Rommel's strategy of forward defence, holding the Allies to their landing areas as much as possible while massing forces for the counterattack that would throw the Allies into the sea. The Normandy *Bocage*—narrow lanes, high impenetrable hedges and small fields—proved excellent defensive terrain. The *Bocage* dominated the invasion area to a depth of up to 50 miles, except on both flanks.

The Allies pressed on, capturing the town of Bayeux and linking all of the beachheads, save that of VII Corps at Utah. Of the 156,000 men landed by sea and air on D-Day itself, approximately 10,000 had become casualties, heavy, but far lower than was expected. New divisions would land every day for the next few weeks as the Allies attempted to build their forces faster than the Germans could contain them.

However, German forces, despite being hampered by continuous Allied air attacks and partisan activity, rushed to seal off the invasion beaches. *12. 'Hitlerjugend' SS-Panzerdivision* (12th 'Hitler Youth' Armoured Division) joined *21. Panzerdivision* in an attempt to restart its D-Day counterattack, but heavy pressure from the British 3rd Division and the Canadian 3rd Division forced them onto the defensive. Nonetheless, their combined strength frustrated all efforts to take Caen. The arrival of *Panzer Lehr Division* (the Armoured Demonstration Division made up of troops used to develop and demonstrate new tactics) at the front completed *I SS-Panzerkorps* (I SS Armoured Corps), but still failed to create enough strength to break through to the beachhead as the British 50th Division fought them to a standstill and forced them on to the defensive as well.

On the American right flank, *709. Infanteriedivision* (709th Infantry Division) and *91. Luftlandedivision* (an army division lightly-equipped for rapid movement by air) continued their battle against the 4th 'Ivy' and 90th 'Tough Ombres' Divisions in the Utah beachhead. Once the location of the landings was clear, *243. Infanteriedivision* moved across from the other side of the Cotentin Peninsula to join them. The veteran *6. Fallschirmjägerregiment* (6th Parachute Regiment) under *Oberst* Von der Heydte sealed off the base of the peninsula, continuing its battle with the American 82nd 'All American' and 101st 'Screaming Eagles' Airborne Divisions.

CARENTAN

Carentan, between the Utah and Omaha beachheads, was, for the moment, the pivotal point in the US sector. While the Germans held it, the beachheads remained separated and could be dealt with one at a time. Rommel ordered *II Fallschirmkorps* (II Parachute Corps) to Carentan with the mobile elements of its *3. Fallschirmjägerdivision* (3rd Parachute Division). There it was joined by *17. 'Götz von Berlichingen' SS-Panzergrenadierdivision* (17th SS Motorised Infantry Division) and the remnants of *352. Infanteriedivision* retreating from Omaha Beach.

The German High Command had committed all of its mobile reserves, but far from pushing the Allies back into the sea, they were barely containing them. Carentan fell at dawn on 12 June, after a night long assault by the 101st Airborne Division supported by naval gunfire and artillery. They held it against counterattacks by *17. SS-Panzergrenadierdivision*, enabling the Americans to link up the Omaha and Utah beachheads. Hitler, meanwhile, issued an order forbidding any withdrawal. The initiative was firmly in Allied hands.

Even while the fight to link the beachheads was being fought, the push for St. Lô was under way. The US V Corps with the 1st 'Big Red One', 2nd 'Indian Head' and 29th 'Blue and Grey' Divisions headed inland from Omaha making good progress as the Germans withdrew to better defensive lines. By 13 June, the defenders, although stretched almost to breaking had halted the Americans short of St. Lô on the Elle River. Their tenacious defence and the difficult bocage terrain had frustrated all efforts at a breakthrough.

OPERATION PERCH—VILLERS BOCAGE

The British 7th Armoured Division, the famed 'Desert Rats', landed their first elements on D-Day. After days of hard fighting, they discovered a widening gap developing between the *Panzer Lehr Division* facing them and the *352. Infanteriedivision* in the US sector on 12 June. The next day an almost unopposed advance through the gap put the lead elements of the 7th Armoured Division into Villers Bocage.

Bold action by a handful of Tiger tanks from *101. SS-schwere Panzerabteilung* (101st SS Heavy Tank Battalion) led by *Obersturmführer* Wittmann, a celebrated Tiger Ace, destroyed the leading companies. With the arrival of *2. Panzerdivision* and counterattacks by *Panzer Lehr*, German pressure forced the isolated 7th Armoured Division to withdraw. After days of continuous heavy German attacks, the Desert Rats returned to British lines on 17 June, its breakthrough pinched out.

CUTTING THE COTENTIN

The US 4th Division's thrust northwards to Cherbourg was proceeding slowly, so on 9 June Bradley changed strategy. Instead of going north, US VII Corps under 'Lightning Joe' Collins would drive hard across the Cotentin Peninsula cutting Cherbourg off from reinforcement. The US 9th 'Old Reliable' Division reached Barneville on the west coast on 17 June, cutting off the remains of *77*, *243*, and *709. Infanteriedivisionen* in the Cotentin Peninsula.

HEDGEROW HELL

On 14 June the First US Army reorganised. A new VIII Corps faced south at the base of the Cotentin Peninsula and the new XIX Corps took over the centre while V Corps remained on the eastern flank adjacent to the British sector. All three corps renewed the thrust southwards on 16 June, but by 20 June their attacks had again been contained after an advance of several miles.

The hedgerows of the *Bocage* were proving incredibly strong defensive terrain. To make matters worse, the badly weakened *352. Infanteriedivision* was reinforced by battlegroups from two more infantry divisions allowing it to continue holding the line, even if only just.

The Great Storm

By 17 June, the Allies had landed 557,000 troops, 81,000 vehicles and 183,000 tons of supplies. They were winning the race to build up their forces faster than the Germans could reinforce theirs. In this they were helped by FUSAG, the fictitious First US Army Group under General Patton. Allied deception measures kept the Germans expecting FUSAG to launch the 'real' invasion in the Pas de Calais, forcing them to hold their reserves back to oppose it. Meanwhile, the German divisions actually committed in Normandy were paying dearly in the attritional battles, exposed to air attack and naval gunfire as they were.

However all was not well. The Overlord plan called for Caen and Cherbourg to be taken by June 23 and St. Lô soon after. So far none of these had yet been taken. Unless the Allies could make faster progress, there was a danger that the Germans could bring in more divisions, sealing the beachhead off, and causing a prolonged stalemate. To make matters worse, the Allied buildup was hindered by a great storm that lashed the invasion beaches from 18 to 21 June. The storm severely damaged the two floating 'Mulberry Harbours' the Allies had built on the landing beaches, disrupting the landing of additional divisions and the stockpiling of supplies, as well as grounding air support.

Cherbourg

The last week of June saw renewed efforts by both Allied armies to capture Cherbourg, St. Lô and Caen. On the western flank the US VII Corps turned to fight its way up the Cotentin Peninsula toward Cherbourg, finally taking the port on 27 June after a week-long battle amongst the ruins of the city. Unfortunately the combination of German demolitions and the battle thoroughly destroyed the port. It was not until the end of September that it reached its full capacity supplying the Allied advance.

Operation Epsom—Hill 112

While Bradley's US First Army besieged Cherbourg, Dempsey's British Second Army made another determined attempt on Caen in what would be the third biggest tank battle of the war, Operation Epsom. On 26 June, the British VIII Corps attacked *12. 'Hitlerjugend' SS-Panzerdivision* in the cornfields west of Caen. The plan was to cross the river Odon and capture Hill 112 before swinging eastwards across the Orne and isolating Caen. Subsidiary attacks by I and XXX Corps a day earlier secured the flanks of the operation.

Pounded by a preparatory bombardment from over 700 guns, the young grenadiers of *Hitlerjugend* fought grimly for every village. Despite counterattacks by *Panzer Lehr*, *2.* and *21. Panzerdivisionen*, and the newly-arrived *1. LSSAH SS-Panzerdivision*, the 11th Armoured Division reached the northern slopes of Hill 112 late on 28 June. The next day heavy counterattacks by *II SS-Panzerkorps*, two more SS armoured divisions transferred from Hungary, threatened the flanks of the operation. Faced with elements of seven armoured divisions, VIII Corps pulled the 11th Armoured Division off Hill 112 on 1 July, ending Operation Epsom.

While failing to take Caen, Epsom did keep the German command focused on their vulnerable right flank. Good tank country beyond Caen meant that any breakthrough here would be catastrophic, so all available armoured forces were committed to the battles around Caen. With little left to reinforce the American sector, they could only hold on and hope on their left flank, trusting the *Bocage* to contain the US First Army.

General Offensive

With Cherbourg taken and Epsom occupying the German reserves, the US First Army turned south to secure St. Lô and prepare for the breakout. The US XIX Corps renewed its offensive on 29 June with an attack on Villiers Fossard by 3rd Armored Division followed by 29th and 30th Divisions pushing towards St. Lô. On 3 July, the US VIII Corps renewed its drive southwards down the western coast of the Cotentin Peninsula towards Coutances, followed the next day by VII Corps now inserted between the VII and XIX

Corps. Progress through the *Bocage* was still frustratingly slow, and casualties heavy throughout the first week of July.

The Germans reacted to the US advance by replacing *Panzer Lehr*, weakened by a month of combat on the British front, with *275. Infanteriedivision*. Thus freed up, *Panzer Lehr* shifted eastwards to join the newly arrived *2. 'Das Reich' SS-Panzerdivision*. On 11 July both divisions attacked west of the Vire River towards Carentan. After their initial surprise, the Americans reacted quickly and halted the German counterattack, before renewing their own attacks.

Operations Charnwood & Jupiter

With all attempts to outflank Caen stymied, the British Second Army launched another effort to take Caen, Operation Charnwood, on 8 July following a massive raid by RAF Bomber Command. The strike caused huge damage to the city and its inhabitants, but the resistance encountered from *12. SS-Panzerdivision* showed the defenders were unbroken. The 3rd Canadian Division of the British I Corps wiped out the defending regiment of *16. Felddivision (L)* (16th Air Force Field Division) and took all of Caen north of the river Orne, but could go no further against the ubiquitous *Hitlerjugend*.

On 10 July, the British VIII Corps launched Operation Jupiter, which retook the northern slopes of Hill 112 at great cost, but the southern slopes remained in German hands.

St. Lô

With such slow progress by both the US and British armies, and a casualty rate that was quickly exhausting their reserves of trained manpower, the threat of a stalemate was looming. Politicians and the Press were demanding a breakthrough. To break the stalemate, major operations were planned for mid-July for the US and British armies. Before they could launch their part of the operation, the Americans had to capture St. Lô, an objective which had eluded them for over a month.

On 11 July the US V Corps launched 2nd Division, supported by 200 guns, at Hill 192 overlooking St. Lô. Bad weather stopped the planned air support, but new tactics for fighting in the *Bocage* allowed the division to push *3. Fallschirmjäger-division* off the hill by nightfall. With Hill 192 cleared, 29th Division attacked on 15 July, finally capturing St. Lô on 18 July.

Operation Goodwood

Both the British and US breakout operations had been planned for 18 July, but delays in taking St. Lô meant that the British operation would have to begin first in order to pin down the German armoured forces while the US First Army completed the capture of St. Lô and prepared their own operation. A British breakout was unlikely in these circumstances, with at least six German armoured divisions and three Tiger heavy tank battalions committed to holding the critical British front. There was little choice though, as holding the German armoured divisions in place was vital if the Americans were to break out.

In the British sector, the complex Operation Goodwood swung into action. A subsidiary attack by XXX Corps and XII Corps, Operation Greenline, began on the night of 15 July, pinning down the three armoured divisions of *II SS-Panzerkorps* around the old Epsom salient and drawing in *1. SS-Panzerdivision* as well. This assault was the prelude to Goodwood proper, which commenced on 18 July, after over a thousand Allied heavy and medium bombers pounded the flanks of the attack. The main punch by the British VIII Corps involved over 750 tanks in three armoured divisions attacking on a narrow front, with 250 more with I Corps and II Canadian Corps on the flanks. This attack was supported by 750 field and naval guns, and fighter-bombers in what would be the second biggest tank battle of the war.

The attack began well with the 11th Armoured Division once again reaching the operation's objective of Bourgébus Ridge. At this point Montgomery made several ill-judged statements to the press overstating the success of the operation. This gave considerable ammunition to his critics when traffic jams in the crowded Orne bridgehead delayed the Guards and 7th Armoured Divisions' attacks, allowing the Germans to counterattack in strength. By the end of the day, three German armoured divisions backed by Tiger tanks and massed 88mm guns had brought the offensive to a halt pushing the British off Bourgébus Ridge.

Although the ridge remained in German hands, Caen had finally been taken. The cost had been light both in men and tanks for the extent of the gains. Only 160 British tanks had been destroyed, while the Germans had lost a similar number. The British tanks were quickly replaced, but the Germans could not make good their losses.

The attacks continued with Operation Spring (conducted by the newly-activated First Canadian Army) forcing the Germans to throw their armoured divisions into the meat grinder once again, with no chance to rest and rebuild. The German armoured divisions preparing to move to the US sector, where a breakout through the paper-thin German line was imminent, were forced to remain in the British sector.

Operation Cobra

The stage was now set for Operation Cobra, the American breakout. Like Goodwood, the plan involved saturation bombing on a narrow front opposite the US VII Corps. This was to be followed with an infantry assault backed by the two strongest armoured divisions in existence, the 2nd 'Hell on Wheels' and 3rd 'Spearhead' Armored Divisions. After a false start on 24 July when the bombing hit the American assault waves, Operation Cobra began on 25 July. While bombing again hit their own infantry, the effect of two days of bombing on *Panzer Lehr* was catastrophic.

The assaulting infantry pressed through the German defences, their success exploited by 2nd Armored Division on 26 July, which broke through into open country on a one-road frontage. With no armoured reserves, the Germans could do nothing to stop the trickle becoming a flood. The US VIII Corps joined the attack capturing Coutances on 28 July. On 30 July, Avranches fell and the Allies were finally out of Normandy. In less than a week, VIII Corps had advanced further than it had in the previous two months of hard fighting.

With FUSAG no longer needed and the US Army about to fight in two different directions at once, 3 August saw General Patton appointed to command a new US Third Army with VIII Corps and the new XV and XX Corps containing the bulk of the US armoured divisions. VIII Corps swept into Brittany isolating the ports of St. Malo, Brest, Lorient and St. Nazaire by 7 August.

Operation Bluecoat

The British had not been idle either. By 30 July, the three armoured divisions of the British Second Army had been transferred west adjacent to the American sector. There they attacked alongside US V Corps towards Vire and Mont Pinçon in Operation Bluecoat. *II SS-Panzerkorps* swung west to block this move, but XXX Corps took Mont Pinçon on 6 August, the same day Vire was taken by the US XIX Corps. The German left flank had completely crumbled. The US Third Army had reached Le Mans and Nantes deep behind their flank on 8 August. Only desperate measures could stop the Allies south of the Seine.

Mortain Counterattack

Hitler, alarmed at the breakthrough, ordered all of the German armoured divisions out of the line to mass for a counterattack that would cut the Third Army from its supplies in Normandy. Insufficient infantry divisions to relieve the armoured units, and constant Allied attacks all along the line, made it impossible to pull most of the armoured divisions out of the line. The troops that could be mustered amounted to little more than strong battalion battlegroups from six different divisions, amounting to no more than 185 tanks in total, less than one full-strength division.

The attack, Operation *Lüttich*, started on the night of 6 August, hitting the US 30th Infantry Division hard around the town of Mortain. With daybreak, Allied air power struck the German assault, as American infantry staunchly held on to key positions. Only limited penetration was made with very heavy losses for the attackers. Despite advice from his generals urging a withdrawal to the Seine River, Hitler ordered the army to hold their ground.

Operations Totalize & Tractable

This played straight into the Allies' hands. While the US Third Army encircled the Germans from the south, their First Army and the British Second Army pushed them eastward. This left the First Canadian Army to close the neck of the pocket that had formed.

The First Canadian Army launched Operation Totalize aimed at Falaise on 8 August. Unusually they attacked with dense columns of tanks in the dark of night, easily smashing through the German front line. Heavy counterattacks by *12. SS-Panzerdivision* halted the advance half way to their objective on 11 August. On 12 August, the US XV Corps reached Argentan south of Falaise. The Canadian offensive resumed on 14 August with Operation Tractable, breaking through to Falaise on 17 August. The concentrated German forces were now squeezed into a pocket, with the Falaise gap the only way out.

The Falaise Pocket

Allied air power took a terrible toll on the Germans. On 16 August, Hitler authorised retreat from the pocket, but it was too late. The Canadians moved to close the gap, succeeding by 20 August, with the 1st Polish Armoured Division taking a crucial role in the bitter fighting. By 22 August, resistance within the pocket had ceased.

The Seine and Paris

While the Falaise Gap was being closed, the US XV Corps had handed Argentan over to the US V Corps and headed east. Its 79th 'Cross of Lorraine' Division was the first across the Seine on 19 August. On 25 August, D+80, the French 2nd Armoured Division had the honour of liberating Paris. By then all four Allied armies had closed up on the Seine River, ten days ahead of the schedule set by General Montgomery's original plan for Operation Overlord.

Of the approximately one million men who fought for Germany (by no means all of them German) during the Normandy campaign, nearly 240,000 were casualties and a further 200,000 prisoners or missing. The cost had been high, but the Normandy Campaign was undoubtedly an Allied victory.

NORMANDY

US forces burst ashore on Utah and Omaha beaches.

Elite British paratroopers silence guns firing on the invasion beaches.

US rangers clear the exits of Omaha beach with Sherman tanks in support.

British commandos clear the flanks of the British beachhead.

AIRBORNE ASSAULTS

After months of training and preparation in England D-Day is finally here! Lead the men of your Airborne Company in a daring airborne attack as they spearhead the invasion of North-west Europe which can be represented with the Seize and Hold Mission.

Forces for Airborne Assaults

Although many of the airborne units of the Allied Airborne divisions have been converted from regular 'ground-pounders', all have received special training and equipment to enable them to undertake airborne operations. Once their airborne mission is complete they will be reinforced by supporting ground-based troops, containing much heavier equipment and landed by sea in Normandy.

Unlike their British counterparts, the US glider rifle companies, although landed from the air, are not trained to conduct airborne assaults. Rather they are delivered to the battlezone a few hours after the main paratrooper assault goes in, with the intention of providing reinforcements to the hard-pressed paratroopers.

Only a British Parachute or Airlanding Company or a US Parachute Rifle Company may conduct an airborne assault using the Seize and Hold mission.

When conducting an airborne assault your force must only contain support platoons that have Parachute, Glider or Airlanding in the title, or Air Support.

Additionally, no Snipers may be included in your force for an airborne assault as every rifleman is expected to engage in close combat!

Prepare your Airborne Force

For months now your planners have been beavering away finalising force composition, aircraft payloads, timetables, and flightpaths. Every soldier and his equipment has to be accounted for in the plan. Because of the dangers of travelling by air to the battlefield, senior commanders were always transported in separate planes or gliders.

Before Deployment in an airborne assault you must assign all HQ support weapons to platoons. No platoons may make combat attachments for an airborne assault.

Each of the Warrior and Independent teams in your force must join a platoon for the purposes of deployment. However, your Company Command team and 2iC Command team must join different platoons for an airborne assault.

Preparation

Before Deployment: Assign HQ support weapons and Join all Warrior and Independent teams to platoons.

Conducting an Airborne Assault

Wind Direction

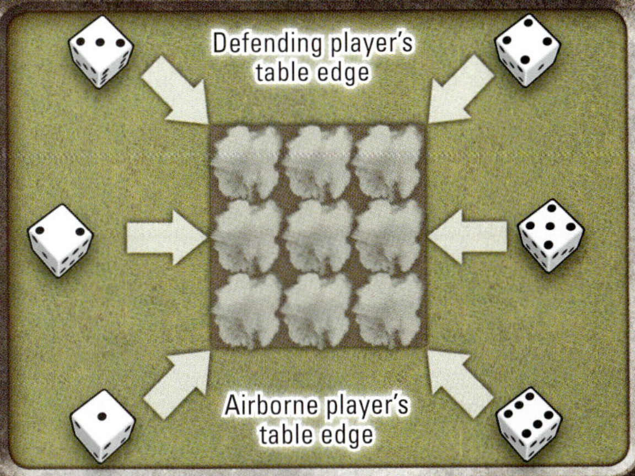

Roll for Wind Direction

Despite the best efforts of meteorologists, wind direction is one of the least predictable aspects of airborne assaults. However, since your parachutes will be scattered by it and your gliders will make their landings in the face of it, the way the wind is blowing has a major impact on your airdrop.

To determine the wind direction, the airborne player chooses either long table edge as their own, then rolls a die and checks the Wind Direction Diagram to determine where the wind is coming from.

PATHFINDERS

A few hours ahead of the main airborne assault force small teams of highly trained paratroopers equipped with lights and radar guidance equipment are dropped on the target areas. The pathfinders quickly find the pre-planned landing and drop zones for the main party and set about marking them with lights and activating their Eureka radar beacons to guide the aircraft in.

During your deployment, choose any table edge or corner and place your Pathfinder marker on it. During the game whenever you roll for Scattered Reserves any die roll of 2 or 5 will result in the platoon arriving on the table edge or corner marked by your Pathfinder marker. If the marker is located on a corner, any platoons arriving there must enter the table within 16"/40cm of the corner.

British pathfinder.

US pathfinder.

PATHFINDERS

Place the Pathfinder marker on any table edge or corner. Platoons will arrive here from Scattered Reserves on a roll of 2 or 5.

PARACHUTE LANDINGS

In preparation for the largest airborne assault in history the Allied airborne commanders have devised new tactics for their parachute forces. Instead of dropping the men 'on the objective' and relying upon the tough paratroopers to form themselves into small groups and launch somewhat haphazard attacks as they are able, drop zones have been carefully selected near the intended objectives. Around each drop zone is a series of rendevous points (RV's) where each paratrooper must report upon landing. As soon as the unit commander has collected enough troops to begin their assault they get underway and make for their objective.

Parachute platoons do not deploy on table at the start of the game, instead each platoon will move on during the game as detailed in the mission rules.

DEPLOYMENT

Parachute platoons do not deploy on table.

CONTAINERS

Heavier weapons and equipment for the paratroopers, like machine-guns and mortars, have to be dropped in special containers and recovered after landing. This was priority number one for the paratroopers.

All Parachute platoons have collected their containers and are fully armed when they move onto the table during the game.

COUP-DE-MAIN GLIDER ASSAULT

When playing a Seize and Hold mission, the Airborne player may use the Coup-de-main rules.

The Airlanding Platoons of the British Airlanding Company were trained to conduct a coup-de-main, an airborne assault aimed directly at vital enemy positions, ahead of the paratrooper landings. Once the enemy position was seized the coup-de-main party would be reinforced by paratroopers as well as gliderborne reserves.

Only the Airlanding Platoons (Combat Platoons) of a British Airlanding Company may make a Coup-de-main Glider Assault. If they do so add +145 points to the cost of each platoon conducting the assault. If they do not, their glider is considered to have landed off-table and the platoon will move onto the table as detailed in the mission rules. All other Airlanding or Glider platoons land off-table and move onto the table as detailed in the mission rules.

Each Coup-de-main platoon is carried in a single Horsa glider. Warrior and Independent teams that have joined the platoon are carried in the platoon's glider for no additional cost.

COUP-DE-MAIN
British Airlanding Platoons may Coup-de-main, if not, they arrive on table as per the mission rules.

CONDUCTING A COUP-DE-MAIN

The Coup-de-main gliders are always at the vanguard of the airborne assault. Unless they are unlucky enough to miss their landing zone.

Attempt to land all Coup-de-main gliders at the start of the first Movement Step.

Each time you attempt to land a Coup-de-main platoon you should complete the following steps for each glider before moving on to the next glider.

FIND THE LANDING ZONE

Several miles from the pre-planned landing zones, the tow planes release your gliders. Out of the blackness of the night sky the gliders swoop towards their objectives. The glider pilots have spent many hours studying aerial reconnaissance photos and maps of the target area. Their keen eyes scan the onrushing ground looking for familiar landmarks to guide them. It is all down to the skill of your pilots to find the landing zone.

For each glider conducting a Coup-de-main, roll a die.

- *On a roll of 2+ the glider finds the Landing Zone.*
- *Otherwise the glider misses the Landing Zone, the glider will land off-table and the platoon it is carrying is now held in Reserve.*

FIND THE LANDING ZONE
Roll 2+ for each glider to find the Landing Zone.

GLIDER LANDING RUN

Having successfully identified the landing zone, the ultimate test for your glider pilots is still to come. They must pick an area as free from obstacles as possible and attempt to land their flimsy wooden flying machine, and its valuable cargo of fighting men, safely.

Pick a landing point anywhere on the table, place the glider on that point facing into the wind. To find the length of the glider's landing run, roll a die. Move the glider forward 2"/5cm for each point on the die roll.

On a landing run roll of 1, the glider crash lands due to pilot error or enemy fire.

OBSTACLES

As the flimsy glider slides along the ground, obstacles strewn across the landing zone, trees, hedges, even enemy fortifications, lurch out of the darkness, threatening to bring the landing run to a premature end.

If the glider's fuselage runs into Rough terrain, an enemy gun (except Man-packed Gun teams), an enemy vehicle or the fuselage of another glider. The glider crash lands and stops immediately. However if the glider's wings hit an obstacle, they just snap off and the wingless fuselage carries on unharmed. The glider only crash lands if the fuselage itself hits an obstacle.

If the glider's fuselage passes through or ends its run on the position of an enemy Infantry or Man-packed Gun team, the defending player moves the team aside just far enough to avoid the body of the glider as they dive out of its path.

GLIDER LANDING RUN

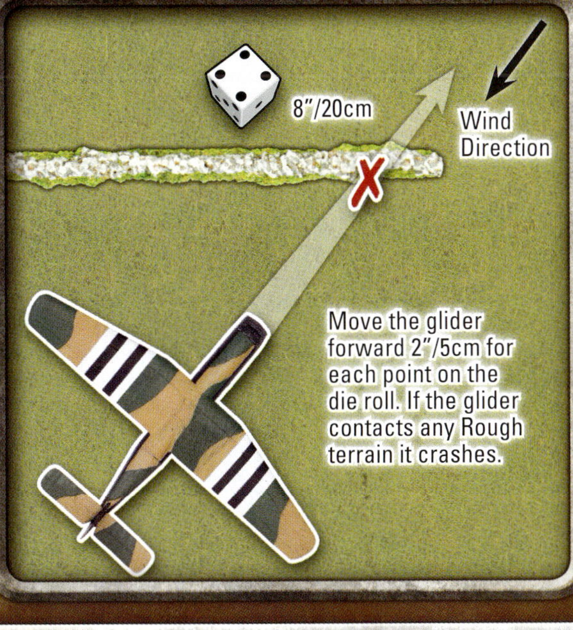

Glider Crash Landings

Glider crash landings are extremely dangerous. The wood and fabric structure of the gliders offers practically no protection in a crash and in fact the splintering framework is a deadly hazard in itself.

If the glider crash lands, the passengers may be killed or injured. Roll a die for each team carried in a crashed glider except Command teams. On a roll of 5+ they survive, but on a lesser roll they are Destroyed.

CRASH LANDINGS

Each passenger, except Command teams, must roll 5+ to survive a crash landing.

After Landing

Dismounting Gliders

Once on the ground, whether their landing was smooth or rough, the Coup-de-main party aren't going to hang around. They realise that speed is key to the success of their mission and the sooner they can get out of this contraption, the sooner they can get at the enemy.

Once the glider has come to a stop, all teams immediately dismount anywhere adjacent to the fuselage of the glider. The platoon may either:

- *move as normal in this Movement Step, or*
- *launch an assault, if they are able, in the Assault Step.*

For this turn only, however, they may only perform one or the other action, not both. All movement is measured from the fuselage of the glider.

Additionally, because they are totally focused on getting amongst the enemy the platoon may not conduct any shooting in the Shooting Step, regardless of whether they move or not.

Dismounting Gliders

Teams dismount adjacent to glider. Coup-de-main platoons may launch an assault or move, but may not shoot this turn.

Shooting Through Gliders

The broken-up fuselage of the flimsy gliders can provide some concealment from, but no real protection against, enemy gunfire.

Gliders are only markers showing where troops land and cannot be shot. Gliders provide Concealment to any teams targeted through them.

Moving Through Gliders

Although gliders are just wood and wires, they are still enough of an obstacle to throw the track of an unwary tank.

A glider counts as Very Difficult Going to teams attempting to move through it.

Rommel's Asparagus

In an attempt to counter airborne assaults, Field Marshal Rommel ordered the erection of thousands of 8 to 12 foot high wooden stakes in possible landing fields all across Normandy. Dubbed *Rommelspargel*, or Rommel's Asparagus, some of these stakes were booby trapped with mines atop them and linked by wire. This ingenious arrangement permitted cattle to graze safely, yet if any gliders attempted to land amongst the stakes they would crash.

The Seize and Hold mission allows the defending player to place three Rommel's Asparagus markers to counter glider landings. Each marker should be modelled 8"/20cm long by 2"/5cm deep. They count as Slow Going, and force any glider that contacts a marker to crash land.

AIRBORNE ASSAULT SUMMARY

STEP ONE: ROLL FOR WIND DIRECTION

- *Choose a table edge, roll a die, and consult Wind Direction Diagram.*

STEP TWO: DROP PATHFINDER

- *Place Pathfinder marker on any table edge or corner. When rolling for Scattered Reserves a roll of 2 or 5 will result in a platoon arriving at this table edge or corner.*

DURING THE GAME:

- *Roll for reserves to bring any Parachute, Glider or Airlanding platoons that are in reserve onto the table.*
- *At the start of the first Movement Step conduct any Coup-de-main glider landings.*
- *In the first Movement Step move that portion of your force that is not in reserve, onto the table.*
- *All platoons arrive on table from Scattered Reserve.*

GLIDER LANDINGS

For Each Glider:

1. Roll to Find Landing Zone

- *Roll 2+ to find Landing Zone, otherwise lands off-table, platoon is in Reserve.*

2. Make Glider Landing Run

- *Place Glider and roll for landing run distance.*
- *Glider crash lands on a roll of 1 or if contacts Rough Terrain, Gun team, vehicle or glider.*
- *Roll 5+ save for teams (except Command teams) in crashed glider.*

3. Dismount Glider

- *All teams immediately dismount adjacent to the glider.*
- *In the first turn Coup-de-main platoons may move in their Movement Step or launch an assault in their Assault Step, not both.*
- *Coup-de-main platoons may not shoot in their first Shooting Step.*

SEIZE AND HOLD MISSION SPECIAL RULES

Night Training

In preparation for their mission, the Allied airborne troops trained extensively at night, and studied maps and aerial photographs of their target area.

The British Parachute Platoon, Airlanding Platoon, and the US Parachute Rifle Platoon may ignore the usual rules regarding movement at night and may instead move normally.

Surprise

Although they may be expecting an airborne assault, the speed with which the attack begins always catches the defenders by surprise. Many of the troops will be away from their positions, eating, sleeping, or otherwise unprepared.

To simulate the time taken to get the defending force fully ready to fight, all defending platoons on the table start the game Pinned Down and all defending armoured vehicles on the table start the game Bailed Out.

Defending teams may not start the game mounted in their transport vehicles or on tanks, and as normal, may not mount Bailed Out transport vehicles or tanks until the crew have remounted.

Airborne Reserves

An airborne commander can't afford to wait around until all of their company have gathered at the rendezvous point. As soon as they have enough troops to begin their mission they head for their objective. The rest will have to follow as soon as they have formed up.

In the Seize and Hold mission both players hold half their platoons in Reserve (see page 268 of the rulebook) (the Attacker ignores any platoons conducting a Coup-de-main). Place these platoons to one side at the start of the game.

At the start of your first turn roll a die. On a roll of 5+ your first reserve platoon arrives. It may be any platoon of your choice, subject to the Unload the Gliders special rule.

At the start of each subsequent turn roll again adding one more die than the previous turn. Each roll of 5+ results in another platoon arriving from your reserves. If you roll three or more dice to get Airborne Reserves in a turn, but fail to score any roll of 5+, you automatically receive a single platoon from Airborne Reserve anyway.

When a reserve platoon arrives place its Platoon Command team at your edge of the table to remind you of its arrival. During your Movement Step move the newly-arrived platoon on from the table edge indicated in the mission briefing.

Unload the Gliders

Gliderborne reinforcements carry heavier equipment like anti-tank guns and vehicles and are slower to unload and get to the battlefield.

During the game all Glider or Airlanding platoons, except those that land on the table in a Coup-de-main, may only arrive on table after all Parachute platoons. Glider or Airlanding Support platoons must arrive last of all.

Airborne Scattered Reserves

In an airborne assault both sides' reserves are scattered over a wide area and could arrive at the front from just about any direction.

When each platoon arrives from Reserve, the player rolls a die to determine from which table edge or corner it will arrive (the mission map shows the edge or corner for each roll). If the reserves arrive from a corner, they must enter the table within 16"/40cm of the corner.

SEIZE AND HOLD

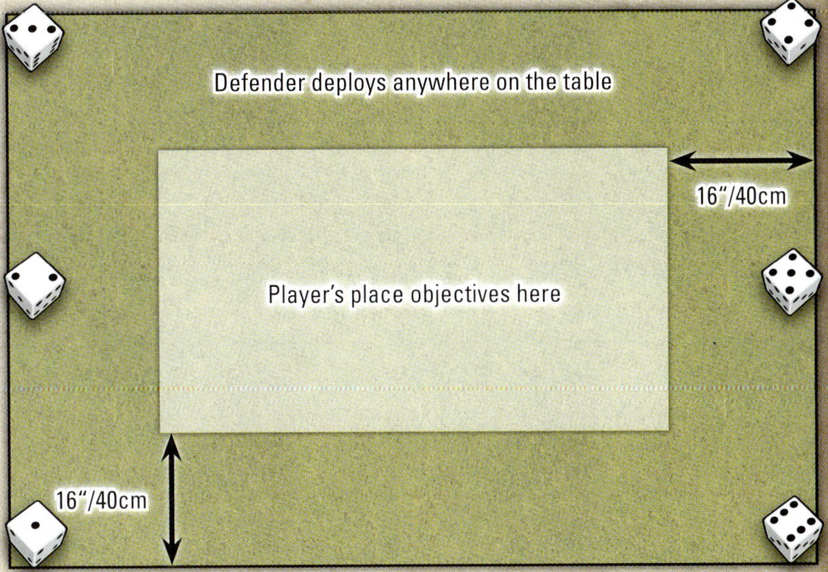

The Seize and Hold mission reflects the daring airborne assaults conducted by the US and British Airborne forces on D-Day.

Mission Special Rules:

Airborne Assault Rules (pages 12-17)
Dawn (page 273 of the rulebook),
Night Training (page 18),
Surprise (page 18),
Reserves (page 268 of the rulebook),
Airborne Reserves (page 18),
Unload the Gliders (page 18),
Airborne Scattered Reserves (page 18),

Your Orders

Attacker

In just a few hours the bulk of your army will be thrown ashore on the beaches of Normandy. To ensure a speedy advance for the ground troops, and to disrupt enemy counterattacks, your force must capture vital positions deep in the enemy rear. You must seize an objective point and hold it until a link-up can be established with your ground troops.

Defender

The enemy has launched his long expected invasion! While your force is far from the beaches, attack from the air is always possible.

Your mission is to hold both objectives and drive the attacking force back to a safe distance.

Preparing for Battle

1. The player with the Allied airborne force is always the attacker. If both players have Allied airborne forces, both players roll a die. The highest scoring player is the attacker.
2. The defender places one objective anywhere on the table at least 16"/40cm from the table edges.
3. The attacking player places two objectives anywhere on the table at least 16"/40cm from the table edges.
4. The defender must hold at least half of their platoons in reserve at the start of the game. The remaining platoons deploy anywhere on the table.
5. The attacker must hold at least half of their platoons in reserve, ignoring any that are to be used in a Coup-de-main. The remaining platoons are not deployed on table at the start of the game and will instead move onto the table in the attacker's first Movement Step using the Airborne Scattered Reserves rule.
6. The attacking player now removes one of the objective markers that they placed (not the objective that the defender placed).
7. The defender may now place up to three Rommel's Asparagus markers anywhere on the table.
8. The attacker selects their table edge and rolls for Wind Direction.
9. The attacker places the Pathfinder marker using the Airborne Assault rules.

Beginning the Battle

1. Neither player may make any Reconnaissance Deployment moves.
2. The attacking player has the first turn. Remember to roll for Reserves in the Starting step.
3. The Defender starts the game in Prepared Positions, all their platoons may begin the game Dug In.
4. At the start of their first Movement Step the attacker lands any Coup-de-main gliders.
5. In their first Movement Step the attacker moves that portion of their force that is not in reserve onto the table as if arriving as Airborne Scattered Reserves. All reserves arrive on the table as Airborne Scattered Reserves.

Ending the Battle

The battle ends when either:

- the attacker starts any turn from turn 6 holding any objective, or
- the defender starts any turn after turn 6 with no attacking teams within 16"/40cm of either objective.

Deciding Who Won

The airborne player wins if they hold any objective. The attack has gained a foothold and reinforcements are on the way.

Otherwise the defender wins. The airborne assault has failed.

Calculate your Victory Points using the Victory Points Table on page 275 of the rulebook.

AMPHIBIOUS ASSAULTS

Each Amphibious assault has its D-Day, whether it was on the shores of North Africa and Italy, or the biggest amphibious assault in history, Normandy. This section allows you to recreate the drama of a small part of a landing operation.

Beach Tables

Understandably, an amphibious landing requires some rather specific terrain—a beach to land on. You can either create a fully-modelled beach or simply use a strip of blue cloth for the sea and a strip of sand-coloured cloth for the beach.

The table needs a small area of sea at the edge of the table giving way to surf crashing on the beach. Between the beach and the hinterland, there are usually sand dunes, a bank or sea wall, cliffs, or marshes. You will need a side table to hold the troops still in their ships off the table as well.

Sea Zone

The sea zone is where landing craft load up and line up for their run in to the beach. It should be about 4-6"/10-15cm deep, just big enough for the landing craft to fit comfortably.

Surf Zone

The surf zone covers the last few metres of sea up to the point where the landing craft beach. It should also be 4-6"/10-15cm deep, again just enough space for the landing craft as they beach to discharge their cargo.

Like the sea zone, the surf zone is Impassable to all teams except amphibious vehicles.

Beach Zone

The beach zone starts from the shallow water at the landing-craft ramp and goes up to the spring high-tide mark where the dunes or sea wall begins.

The beach will usually be 8"/20cm deep and is Difficult Going. The landing-craft obstacles that litter the beach zone do not hinder the movement of teams once ashore and do not provide any cover or concealment for troops on the beach.

Sea Wall

Above the high water mark there is usually some form of bank or sea wall with shingle washed up against it.

A bank is Very Difficult Going. A sea wall is an Anti-tank Obstacle (see page 231 of the rulebook) that provides Bulletproof Cover to Infantry and Man-packed Gun teams sheltering on the seaward side.

Beach Exits

The landing zones were, naturally enough, chosen for their accessible exits. There is little point in landing somewhere were you need to be a commando to get off the beach!

The beach should have at least one exit through the Sea Wall Zone that is Difficult Going.

Hinterland Zone

Once off the beach, the countryside quickly assumes the normal rural Normandy character. Generally the coastal areas are more open than the inland areas with larger fields and fewer houses and woods.

See our website, *www.FlamesOfWar.com* for how to make a beach table.

SETTING UP

Amphibious landings require a little more work than most scenarios. One of the trickiest parts of managing an amphibious assault is making sure that the right troops land at the right place at the right time. Don't worry if you don't succeed with this—the real Generals didn't always manage to coordinate their landings either!

The first steps in making your landing are to select the platoons that will land in the first wave and determine the direction the current is flowing across the beach.

SELECT THE FIRST WAVE

There are never enough landing craft to put all of the troops available ashore at once. You need to decide which units you want to land in the first wave and which will be left in their ships for the follow-up waves.

Troops left at sea will land as soon as the landing craft carrying the first wave have beached, unloaded, and returned to the ships at sea to reload.

Before the game begins you need to select up to half of your platoons as your first wave. Each of these platoons is loaded in a landing craft. Place the platoon's landing craft in the sea zone opposite the place they are to come ashore (bear in mind that they will drift with the current).

All other platoons remain in their ships at sea and are known as the Floating Reserve.

Warrior and Independent teams do not have their own landing craft. They are attached to platoons in the first wave and travel ashore in their landing craft.

Many platoons include trucks and other unarmoured transport that will have difficulty crossing the beach. If these land with the troops they simply present the enemy with easy targets. Platoons may leave their transport behind and not deploy it at all when making an amphibious landing.

CURRENT DIRECTION

Before the game begins you will need to know which way the current is running. This is important since your landing craft will drift off course in a heavy current.

Roll one die and consult the Current Table to determine the direction the current is running.

CURRENT

Die Roll	Fate
1 to 3	Current runs leftward when facing the beach
4 to 6	Current runs rightward when facing the beach

COMING ASHORE

The current is running strongly and the beach obstacles limit their choices, so coxswains have difficulties landing exactly where they want to. Most end up somewhere down current of where they intended to land. Some fall victim to beach obstacles or defending artillery. Others go completely astray in the confusion, only locating the right beach much later.

Once the landing craft are in place, roll a die for each landing craft to see how far it drifted off course during the run in, or if it even made it to the beach. The Landing Craft Table gives the distance down current the landing craft drifts before beaching in the surf zone, or their fate if they do not make it safely ashore.

If another landing craft already occupies the place a landing craft drifts to, it drifts further down current to the first available space in the surf zone. If the landing craft drifts off the table it returns to sea and attempts to land again next turn as if it had been delayed.

Landing Craft

Die Roll	Result
1	Drift 4"/10cm down current.
2	Drift 8"/20cm down current.
3	Drift 12"/30cm down current.
4	Drift 16"/40cm down current.
5 or 6	Delayed. The landing craft was unable to find a gap through the beach obstacles and does not beach. Place the loaded landing craft back in the sea zone.

Delayed Landing Craft

In the pre-dawn darkness and the confusion of hundreds of craft milling around and making for the beach, it is easy for a coxswain to get separated from their flotilla. At the same time beach obstacles and artillery take their toll of landing craft, requiring replacements from following waves to be hastened into action.

Delayed landing craft return to the sea zone carrying their load and attempt to land again next turn.

This may represent either landing craft being unable to beach or the delay as landing craft destroyed by the enemy defences are replaced with follow-on troops.

DISEMBARKING

As soon as the landing craft ramp goes down, the troops aboard rush ashore, eager to get across the killing zone as fast as possible.

Passengers and vehicles disembark from a landing craft during the Movement Step. They begin their movement from the landing craft's ramp as if they had occupied the same position as the landing craft.

Remember, the Beach Zone is Difficult Going. If a vehicle Bogs Down when disembarking, place it on the beach adjacent to the landing craft's ramp (leaving room for following vehicles to disembark) and mark it as Bogged Down.

Landing craft cannot shoot or be shot at, and take no part in assaults.

Coming Ashore

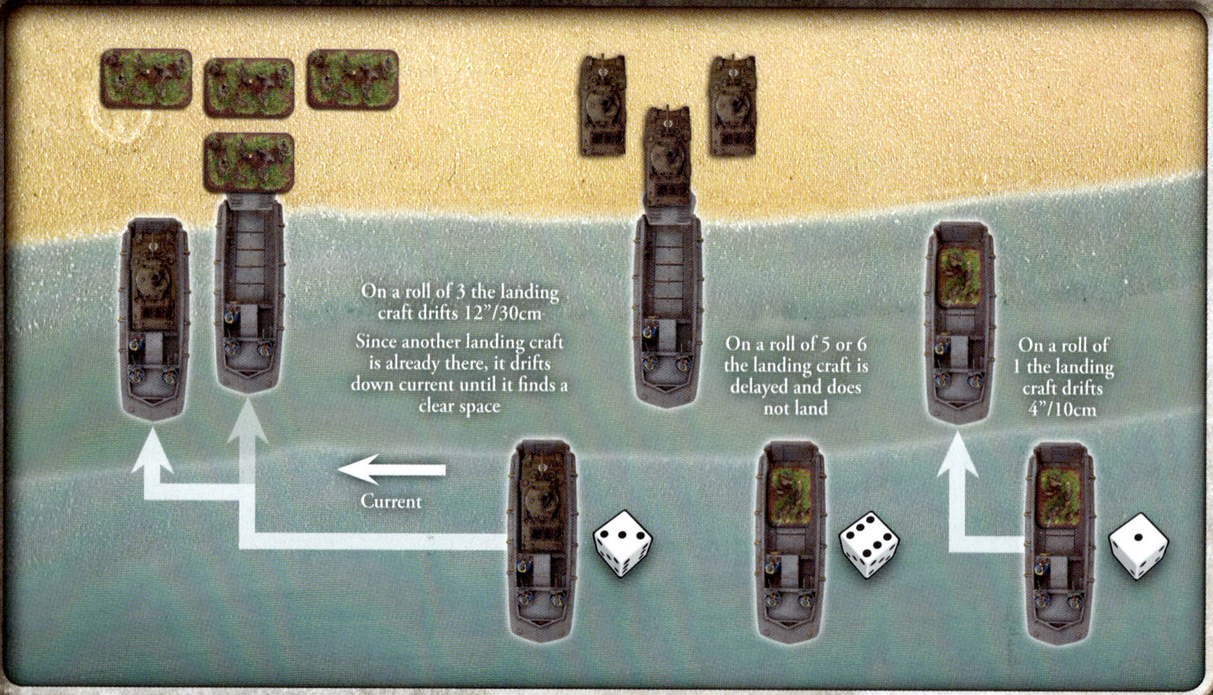

Disembarking

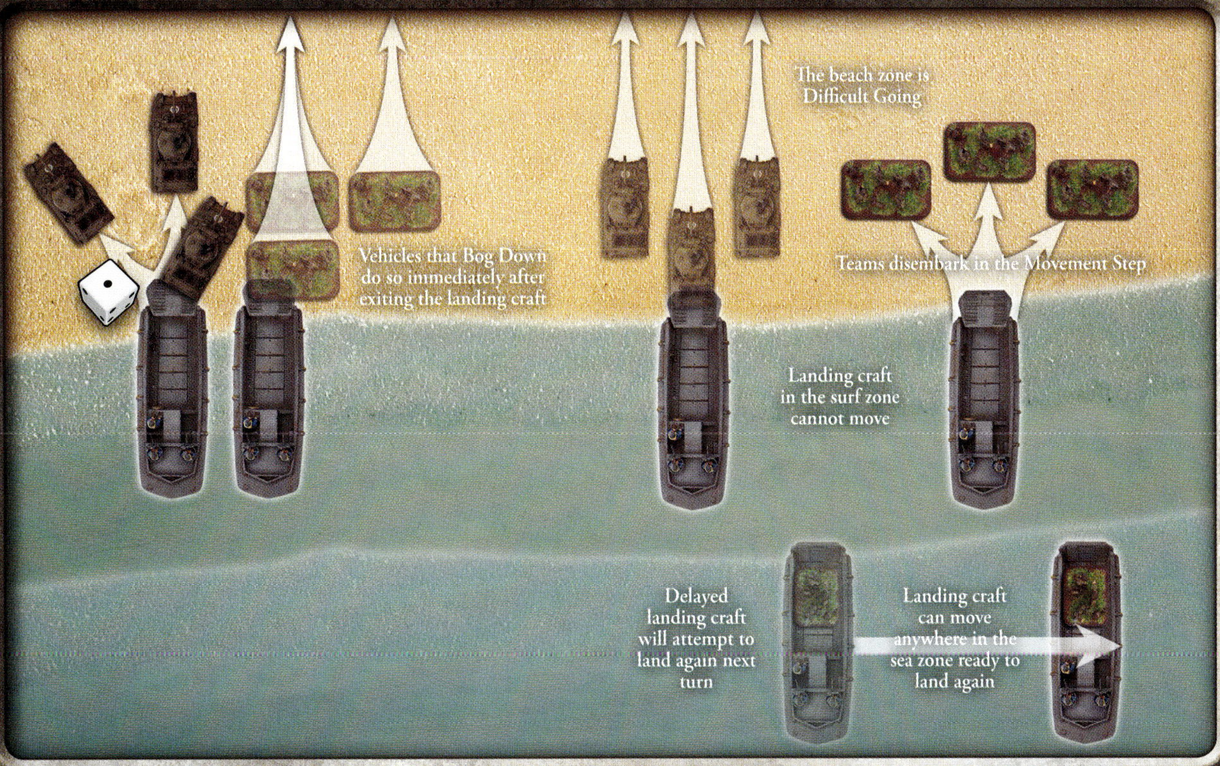

RETURNING FOR MORE

Once unloaded, the landing craft quickly head back to their ships to reload with follow-up troops and return with the next wave to land.

In the same Movement Step after unloading, empty landing craft may attempt to return to sea and reload. Roll a die for each empty landing craft in the surf zone.

- *On a roll of 4+, the landing craft refloats and returns to the Floating Reserve for reloading. Select a platoon from the Floating Reserve to load and place the landing craft in the sea zone to land again next turn.*
- *On any lesser roll, the craft is stuck fast and must try to return to sea again next turn.*

Returning For More

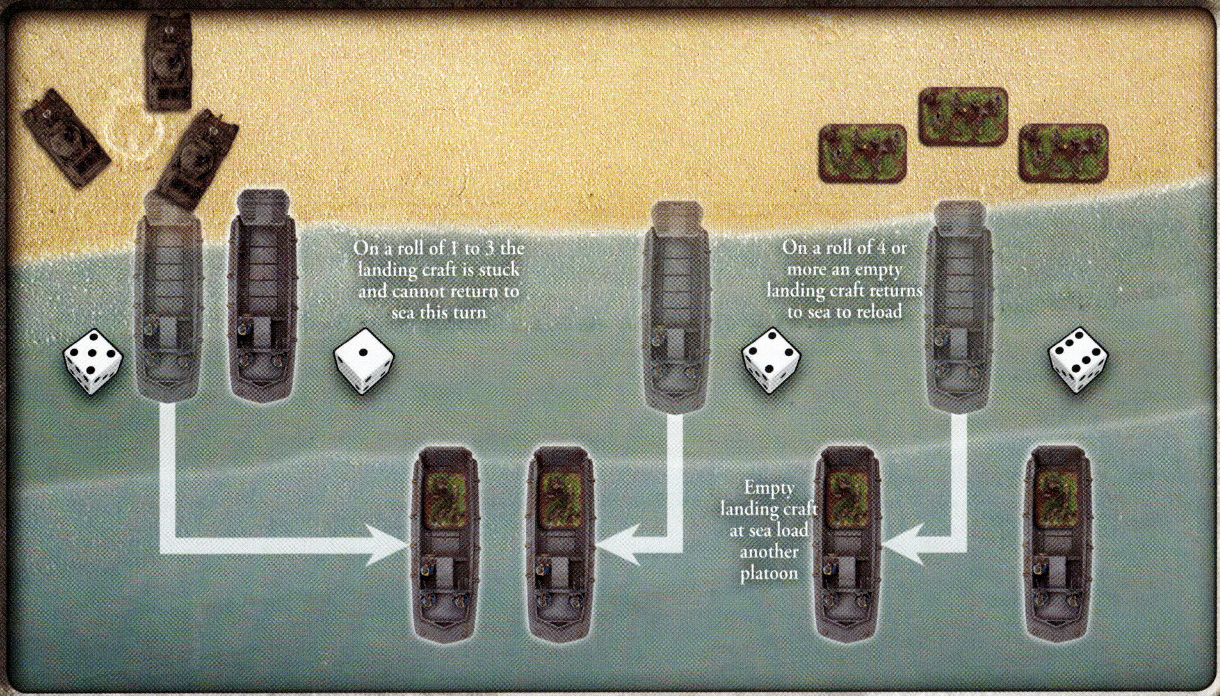

DD TANKS

One of the more unusual weapons landing on D-Day is the Duplex Drive or DD tank (unsurprisingly nicknamed the 'Donald Duck' tank in the US Army!). This is a Sherman tank fitted with a collapsible canvas screen around the top hull and propellers to allow it to swim. Unfortunately, its low freeboard means that it is vulnerable to being swamped by large waves, making its use risky in stormy seas.

A British Armoured Platoon or US Tank Platoon equipped solely with Sherman tanks armed with 75mm guns may fit them out as DD tanks at no extra cost. The platoon retains all of its previous ratings. However, Sherman DD tanks cannot use their hull MG due to the collapsed screen covering its position.

COMING ASHORE

The DD tanks are scheduled to land ten minutes before the first wave of landing craft. Sitting at the water's edge, they will give covering fire as the infantry disembark and then help them deal with bunkers and defences as they force their way inland.

DD tanks do not require landing craft as they can swim ashore on their own ahead of the first wave of landing craft. Because of this, you would not normally select a DD tank platoon to be in the half of your force initially loaded into landing craft.

However, you can bring your DD tanks ashore in landing craft as normal tanks, landing 'dry shod' as the expression goes. If you want to do this, have them occupy one of your landing craft instead of swimming them ashore.

Before the game begins place DD tanks in their swimming form in the sea zone opposite the place they wish to land (bear in mind that they will drift with the current). Roll a die for each DD tank and consult the DD Tanks Table to see how far it drifted off course beaching in the Surf Zone, or if it even made it ashore.

DD TANKS

Die Roll	Fate
1	Drift 6"/15cm down current.
2	Drift 12"/30cm down current.
3	Drift 18"/45cm down current.
4	Swamped. The DD tank was swamped by heavy waves on the run in. It is Destroyed and removed from the game.
5 or 6	Delayed. DD tanks are not fast swimmers. This tank has not yet reached the beach. Place the tank back in the sea zone to attempt to land again next turn.

FIGHTING ASHORE

Since DD tanks can land before the first landing craft, they are ready and able to provide fire support for the first wave as it comes ashore. This extra fire support can be vital in the crucial first few minutes of the landing. Some DD tanks left the rear of their screen erected on landing and remained in the surf giving fire support. Others moved up the beach to engage targets they could not reach from the water's edge.

Once ashore DD tanks may either remain in the Surf Zone providing fire support for the attacking infantry, or move on to the beach as normal.

DD tanks that remain in the Surf Zone do not count as moving and may shoot at their full rate of fire, but must roll a die in each Movement Step.

- *On a roll of 1, the rising tide washes over the tank, which is Destroyed.*
- *Otherwise the tank is fine and can keep shooting.*

COMMAND AND CONTROL

DD tank platoons are very likely to come ashore scattered and piecemeal making it difficult for the platoon commander to coordinate their actions. As a result they are trained to and expect to operate independently.

DD tank platoons that swim ashore (as opposed to arriving by landing craft) operate as Independent teams.

FLOATING ARTILLERY

Both the British and US armies provided the assault divisions with self-propelled guns mounted in landing craft to give fire support as the division assaulted the beaches.

You may load a landing craft with a British Assault Field Battery (SP), Royal Artillery or a US Armored Field Artillery Battery and hold it out to sea firing as artillery rather than bringing it ashore. If staging an amphibious assault US players may field one Armoured Field Artillery Battery in place of a normal Field Artillery Battery, and British players may field one Assault Field Battery (SP), Royal Artillery in place of a normal Field Battery, Royal Artillery (see Turning Tide*).*

If you do so, the landing craft remains in the sea zone rather than attempting to beach, but the artillery on board may fire Artillery Bombardments while still in the landing craft. While at sea the battery cannot be shot at. You may elect to beach the landing craft and bring the battery ashore in a later turn when the beach is secure enough.

The artillery may either spot targets for themselves from the landing craft, or have their normal spotters and observer teams locate targets for them. Measure the range to the target from the landing craft.

Batteries firing from landing craft cannot fire Smoke Bombardments. Due to the constant motion of the landing craft Staff teams cannot operate effectively, so the artillery batteries using the Floating Artillery special do not count as having a Staff team. This means they cannot use rules such as All Guns Repeat! to repeat bombardments, nor may they fire a Time On Target bombardment.

DUKW AMPHIBIOUS TRUCKS

As well as landing craft and DD tanks, the Allied forces landing at Normandy had numbers of DUKW amphibious trucks available. Most of these were used to unload supplies and carry them directly inland, but a small number were used to bring ashore artillery pieces getting them into action more quickly than landing craft would have.

Any US Cannon Platoon or Field Artillery Battery may replace its 1½ or 2½-ton trucks with DUKW amphibious trucks at no extra cost. The command and staff teams ride in one of the DUKW trucks, leaving their vehicles behind. A DUKW amphibious truck counts as a 2½-ton truck with the addition that it is amphibious.

A platoon mounted in amphibious DUKW trucks does not require a landing craft to come ashore. It can swim ashore from the Floating Reserve on its own as if each DUKW truck were a landing craft, except that they use the DD Tanks Table rather than the Landing Craft Table to determine their fate during the run ashore.

Once landed, DUKW trucks simply drive up the beach to wherever they are intended to unload and set the guns down.

DUKW trucks carry their gun in their cargo compartment, but are not strong enough to allow the gun to fire while aboard. The Gun team must dismount beside the truck as if it were unlimbering before it can fire. Since the DUKW cannot tow the gun, nor reload it in the cargo compartment, the DUKW truck is immediately Sent to the Rear when the gun dismounts and cannot be Brought Forward (see page 48 of the rulebook).

HIT THE BEACH NAVAL GUNFIRE SUPPORT

The D-Day landings were supported by dozens of warships from mighty battleships like the *USS Texas* and *HMS Rodney*, down to lowly destroyers. Many of these warships were tasked with providing fire support to the troops on the beaches.

A force making an amphibious landing may have Naval Gunfire Support from warships off shore at a cost of 200 points.

Naval Gunfire Support

If you have Naval Gunfire Support, your force will field an Observer Rifle team that can spot for an artillery battery of four Confident Trained Naval Guns with no Staff team. The Guns are not deployed on the table, but have the range to hit any target on the table.

Weapon	Range	ROF	Anti-Tank	Firepower
Naval Guns	112"/280cm	-	6	1+

HIT THE BEACH MISSION SPECIAL RULES

D-Day Fortifications

The German defences on the coastline can be formidable. They can include bunkers with the latest and heaviest anti-tank guns as well as emplaced tank turrets.

The defending force is best represented by a German Festungskompanie from page 22 of *Earth & Steel*. You may like to defend in a Hit the Beach mission with another type of Fortified Company, though not all of these are ideally suited to defending a beach front.

A Defending force in a Hit the Beach mission counts as a Fortified Company. In particular, remember the Armoured Reserves special rule on page 269 of the rulebook.

Preliminary Bombardment

Before a major attack the defender is subjected to a preliminary bombardment from every heavy gun, bomber, and battleship available in order to pin the enemy down and weaken defenders before the battle begins.

To conduct a preliminary bombardment, the attacking player works through the defending platoons deployed on the table one at a time, including platoons that are held in Ambush.

The attacker rolls a die per team in the platoon as if the whole platoon were under the template of an artillery bombardment. For each roll of 4+ the defending platoon takes a hit on a team in the platoon chosen by the defender. Hits must be assigned as though they were caused by an Artillery Bombardment (see page 132 of the rulebook).

The defender then rolls a save for each Infantry or Gun team that was hit. Any team that fails their save is removed from the game even if it is in Bulletproof Cover.

The defender gets an Armour Save roll for armoured vehicles. The Anti-tank rating of the heavy guns firing the bombardment is 6, so if the die roll plus the vehicle's Top armour is 5 or less, the vehicle is removed from the game. On a higher roll the vehicle is unharmed.

Vehicles in Tank Pits roll their extra 4+ Tank Pit Save if they fail their normal Armour Save. If they pass this, they are unharmed and remain on table.

Unarmoured vehicles do not have any save due to the intensity of the bombardment and are automatically removed from the table unless they are in a Tank Pit where they get their 4+ Tank Pit Save.

To reflect the mind-numbing fury of the bombardment, all defending platoons on the table start the game Pinned Down and all defending Armoured vehicles on the table start the game Bailed Out.

Defending teams may not start the game mounted in their transport vehicles or on tanks and, as normal, may not mount up Bailed Out transport vehicles or tanks until the crew have remounted.

Company and Higher Command teams and other Warrior teams are hardened veterans and always survive a preliminary bombardment, so do not roll to hit them.

Remember that losses taken before the first turn of play do not affect Platoon and Company Morale Checks.

Interdiction

The Allied air forces totally dominate the skies over the landings and far inland. The German reserves are placed centrally so they can respond to a landing anywhere. However, their march is continually interrupted by air attacks making their movement very slow and uncertain.

The Defending player's Delayed Reserves arrive on rolls of 6 rather than 5+ as normal.

Overwhelming Force

Amphibious assaults require overwhelming force. The troops land in the teeth of the enemy defences, scattered and disorganised. It is only through overwhelming force that they will secure a lodgement.

The Attacking force has 25% more troops than the Defending force. A good force for most tables would be 1500 points for the Attacker against 1200 points for the Defender.

When an Attacking platoon is Destroyed, the player returns it to the Floating Reserve to represent a fresh unit landing in its place. Infantry and Man-packed Gun teams in the platoon return to the Floating Reserve and can land again as long as the platoon contained at least four such teams at the start of the game. However, all other types of teams are permanently lost. As usual, Destroyed Armoured vehicles remain where they are.

Warrior and Independent teams are also permanently lost when they are Destroyed.

The Attacking player may voluntarily destroy platoons and return them to the Floating Reserve in the first part of the Starting Step.

No Retreat, No Surrender

The troops landing on the beaches know that it is do or die time. There is no going back, only death or victory. Likewise the defenders have no contact with their headquarters and have no option but to fight on.

Neither side needs to take Company Morale Checks. They will fight until the end.

Total Air Superiority

This battle is the focus of the attacker's strategic plan. Their air force is working overtime attacking enemy airfields and flying extra fighter patrols to keep the defender's air forces out of the battle.

The defending player must roll a 6, instead of 5+, in order to receive ground-attack air support.

Victory Points Table

Outcome	*Results*	*Attacker's Points*	*Defender's Points*
Attacking player holds both objectives by turn 9	Stunning Attacker's Victory	6	1
Attacking player holds both objectives	Major Attacker's Victory	5	2
Attacking player holds the objective placed by the Defending player	Minor Attacker's Victory	4	3
Attacking player holds the objective placed by the Attacking player	Minor Defender's Victory	3	4
Attacking player holds no objectives, but has troops in the Hinterland Zone	Major Defender's Victory	2	5
Attacking player holds no objectives and has no troops in the Hinterland Zone	Stunning Defender's Victory	1	6

HIT THE BEACH MISSION

The time has come to liberate Europe from the Nazi menace. Your force is landing in the first wave. Can you break through Hitler's Atlantic Wall and make the invasion stick?

Mission Special Rules:

Amphibious Assaults Rules (pages 20 to 25),
D-day Fortifications (page 26),
Delayed Reserves (page 269 of the rulebook),
Interdiction (page 27),
Overwhelming Force (page 28),
No Retreat, No Surrender (page 28),
Preliminary Bombardment (page 26),
Prepared Positions (page 264 of the rulebook), and
Total Air Superiority (page 28).

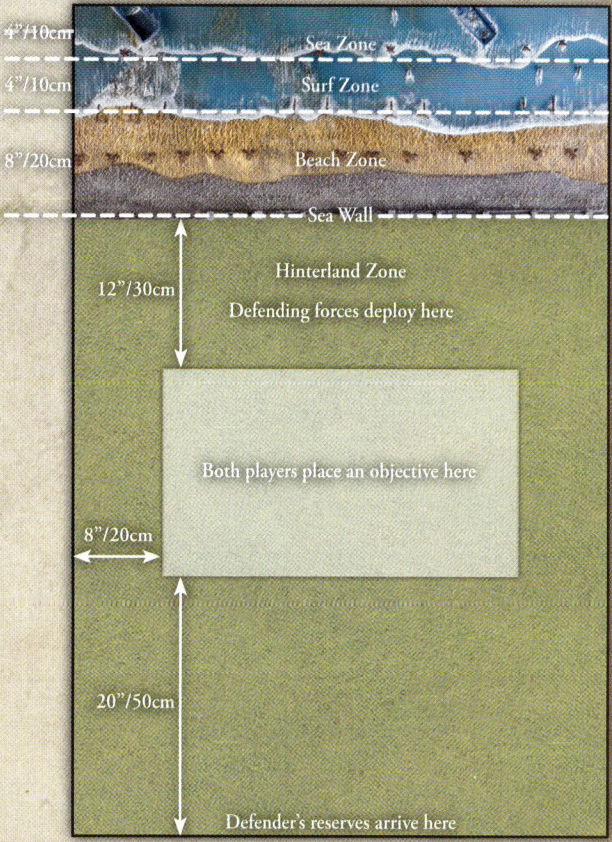

Your Orders

Attacker

Your troops are the first wave ashore. You must push inland and secure the beachhead at any cost. It is essential to capture both objectives as soon as possible.

Defender

The enemy invasion has finally begun. Under no circumstances can they be allowed to gain a beachhead. They must be destroyed on the beaches. You must hold both objectives.

Preparing For Battle

1. The Defending player places their objective first, then the Attacking player places theirs. Objectives cannot be placed within 12"/30cm of the beach or dunes or within 20"/50cm of the rear of the table, nor within 8"/20cm of the table edges.
2. Next the Defending player places their fortifications anywhere in the Hinterland Zone. Bunkers may not be placed across roads or the beach exit.
3. The Defending player must hold at least half of their platoons in Delayed Reserve at the start of the game. The remaining platoons and Independent teams are now placed in the Hinterland Zone.
4. The Attacking player must hold at least half of their platoons in their Floating Reserve. Place the rest in their landing craft in the Sea Zone.
5. Next the Attacking player places any DD tank platoons that will swim ashore, and any artillery batteries carried in DUKW amphibious trucks, in the Sea Zone.

Beginning The Battle

1. The Defending force begins the game in prepared positions, so their troops are in Foxholes and Gone to Ground.
2. The Attacking Player conducts a Preliminary Bombardment on the defender's forces.
3. The Attacking player then lands their DD tank platoons.
4. The Defending player may not make Reconnaissance Deployment moves.
5. Both players should remember that the Defender's reserves will be hindered by the Interdiction special rule when making their plans.
6. The Attacking player has the first turn. They may not make a Night Attack.

Ending The Battle

The battle ends either:

- at the end of the Starting Step of the Attacker's turn 12, or
- when the Attacking player starts any of their turns holding both objectives.

Deciding Who Won

Unlike most missions, casualties are irrelevant. All that matters is getting ashore as far and as fast as possible, or stopping the enemy from doing so.

The Attacking player wins if they hold the objective placed by the Defending player, otherwise the Defending player wins.

Calculate your Victory Points from the Victory Point Table on the previous page.

1

THE BATTLE FOR UTAH BEACH

A navigational error has landed the US amphibious assault force on the wrong sector of Utah Beach. Will this spell doom for the landings? Or will it be a blessing in disguise?

H-Hour 0630 hours: Following a heavy naval and aerial bombardment the first assault waves of the 8th RCT from the US 4th 'Ivy' Division hit the beach at Utah. The landings were timed to take advantage of a low tide exposing the German beach obstacles. First ashore were Companies B and C under Lieutenant Colonel Conrad Simmons, at the sector code-named Tare Green, and E and F under Lieutenant Colonel Carlton MacNeeley, at Uncle Red. The Assistant Divisional Commander Brigadier-General Theodore Roosevelt, son of the former US President of the same name, landed in the first wave. He would be awarded the Medal of Honor for his part in the assault that day. A few minutes after the first infantry landings, Sherman DD Tanks of the 70th Tank Battalion arrived on the beach.

Once ashore the assault troops discovered that rough seas and a strong current had pushed them some 2000 metres (1.24 miles) south of the intended landing site at les Dunes de Varraville. Luckily, this area of the beach, around La Grande Dune, was actually more lightly defended than the intended landing site. The German resistance nest in the area, WN 5, was manned by a single platoon of *919. Grenadierregiment* under *Leutnant* Jahnke, a veteran of the Eastern Front. The preliminary bombardment had severely damaged the German defences, knocking out most of their anti-tank guns, and the US infantry and tanks quickly overwhelmed the position.

The assault troops moved inland and captured the fortified hamlet at La Madeleine, guarding the entrance to the causeway known as Exit 2. The causeway crossed marshland deliberately flooded by the Germans to bar the way inland for an invasion force. At 0900 hours troops began moving inland, some of the infantry wading through the marshes. After engineers had cleared the German minefields, four Sherman DD Tanks led the advance down the causeway. The first Sherman foundered on a demolished section of the road and a German anti-tank gun knocked out the second tank, but by 1100 hours the advance had linked up with paratroopers of the 101st Airborne who had seized the other end of the causeway.

The landings had been successful, with remarkably light casualties for the attacking Americans. On D-Day, more than 20,000 troops and 1700 vehicles along with thousands of tons of supplies were landed at Utah Beach.

REFIGHTING UTAH BEACH

While Utah Beach was the most lightly defended of the invasion beaches, the speed with which the US troop landing there got off the beach and through the marshes beyond is a testament to their courage and training. Use the Hit the Beach mission on page 29 to refight this battle in conjunction with the table layout on page 31 that shows the salient features of Utah beach, and the list on page 32 giving the forces from the 4th Infantry Division and *919. Hessian-Thüringian Grenadierregiment* that fought there. To reflect the unique circumstances of the battle for Utah Beach, the following special rule applies.

Preliminary Air Strike

The Allied air forces had the skies all to themselves on D-Day, although the rough weather and low cloud conspired against them. However, at Utah Beach, American aircraft followed up the devastating naval bombardment with a highly effective air strike that wreaked further havoc on the defenders.

The US player may make one free air strike by a flight of one to three (roll an Aircraft die) Ground-attack Aircraft after deployment, but before the game begins.

BEACH DEFENCES

The defences of Strongpoint 5 (*Wiederstandnest 5* or *WN 5* in German) are much weaker than those further along the beach at Strongpoint 9. A type H667 casemate housing a 5cm KwK gun is positioned to enfilade the beach. At the front centre of the strongpoint is an old Renault FT-17 tank turret mounted on a concrete bunker. Two MG42 heavy machine-guns in gun pits facing the sea sit alongside an old French 47mm SA-37 anti-tank gun (Range 24"/60cm, ROF3, AT9, FP4+) and a 75mm mle 1897 field gun (Range 24"/60cm, ROF2, AT8, FP3+). Communication trenches span the defences. The resistance nest is ringed with barbed wire, and a minefield guards the rear of the position.

EXIT 2

Inland from WN 5 is the entrance to Exit 2, a causeway across the marshes. This is guarded by WN 7, a resistance nest set up around the little hamlet of La Madeleine. At the southeastern end of the hamlet covering the causeway, the Germans have mounted a pair of MG42 heavy machine guns, another 75mm field gun, and a Russian 7.62cm PaK36(r). Barbed wire guards the perimeter of the resistance nest. The building shown is the chapel of La Madeleine.

A huge crater blocks the causeway. This, like the marshes surrounding it, is rated as Very Difficult Going in *Flames Of War*. It is impossible to entrench in the marshes.

US 4th Infantry Division

Charlie Company, 1/8th RCT	Assault Company (Confident Trained) (page 172 Turning Tide)
Company HQ	Company HQ
1st Rifle Platoon	Boat Section
2nd Rifle Platoon	Boat Section
3rd Rifle Platoon	Boat Section
Weapons Platoon	Boat Section
3rd Platoon, Dog Company	Mortar Platoon (with three Mortar Sections)
1st Platoon, Baker Company, 237th Engineer Combat Battalion	Assault Engineer Combat Platoon (with two Operating Squads with M4 Sherman Dozer tank)
2nd Platoon, Able Company, 65th Armoured Field Artillery Battalion	Corps Armoured Field Artillery Battery (with one Gun Section)
1st Platoon, Able Company, 70th Tank Battalion	DD Tank Platoon (with five Sherman DD tanks)
395th 'Panzer Dusters' Fighter Squadron, USAAF	Priority Air Support – P47 Thunderbolt

The first wave ashore on Utah Beach consisted of troops from the 8th Regimental Combat Team (RCT) of the US 4th 'Ivy' Infantry Division. During the run in to the beach artillery support was provided by M7 Priest HMC self-propelled guns of 65th Armoured Field Artillery Battalion firing from their landing craft. As the infantry landed, the amphibious Sherman DD tanks of the 70th Tank Battalion came ashore to give immediate tank support.

German 709. InfanterieDivision

3./919. Grenadierregiment	Festungskompanie (Confident Trained) (page 22 Earth & Steel)
2. Grenadierzug, 3./919 GR	Festungs Grenadier Platoon (at full strength, at A)
1. Maschinegewehrzug 4./919 GR	Festungs Machine-gun Platoon (with two HMG Nests, one 5cm PaK38 and one 7.5cm PaK97/38 attached, at A)
3. Panzerjägerzug 919 GR	Festungs Anti-tank Gun Platoon (with one 7.5cm PaK97/38 and one 7.62cm PaK36(r), two HMG Nests attached, at B)
Alarmeinheiten	**Reserves**
Kompaniegruppe	Company HQ (attached to 1. Grenadierzug)
1. Grenadierzug, 3./919 GR	Festungs Grenadier Platoon (at full strength)
3. Grenadierzug, 3./919 GR	Festungs Grenadier Platoon (at full strength)
1. Pionierzug, 919 GR	Grenadier Pioneer Platoon (at full strength) (page 133 *E&S*)
2. Granatwerferzug, 4./919 GR	Festungs Mortar Platoon (with three Mortar Sections)

The German defenders were from *III. Bataillon, 919. Grenadierregiment* of *709. Infanteriedivision*. A platoon of infantry held the WN 5 resistance nest. They were supported by anti-tank guns in a wide variety of calibres positioned in WN 5 and further back at WN 7 guarding the entrance to the causeway. A few platoons of grenadiers and pioneers supported by mortars, formed the local counterattack reserve.

What If Battles

While the forces shown above represent the forces that actually landed on Utah Beach and those that opposed them, you can also play the scenario with different forces to see how the battle would have turned out if the commanders had committed different forces to the battle.

An Allied force attacking at Utah Beach should consist of 1500 points based on an Infantry Company.

An Axis force should consist of 1000 points based on an Infantry Company. Any tank or self-propelled gun platoons should be the last to appear from reserve. Remember to retain all the fortifications on the map on page 29.

2
THE BATTLE FOR SWORD BEACH

Will the British 'Funnies' give the men of Yorkshire the edge they need to overcome the tough German fortifications?

Sword Beach was on the eastern flank of the D-Day landings just west of the Orne River. The British 3rd Division, General Montgomery's own division in the Battle of France in 1940, was assigned the task of landing on Sword Beach and fighting its way inland to link up with the airborne operations across the Orne River and, if possible, capture Caen.

An off-shore reef limited the assault to a single brigade frontage at *La Brèche*, 'the Breach'. The divisional commander, Major-General Tom Rennie, selected 8 Brigade as his assault force with Lieutenant-Colonel C F Hutchinson's 2nd Battalion, The East Yorkshire Regiment landing on Queen Red (as the eastern sector was called). The 1st Battalion, The South Lancashire Regiment was to land on Queen White to the west. The reserve battalion landing behind them as the second wave was the 1st Battalion, The Suffolk Regiment.

The brigade was well supported with specialised assault armour from the 79th Armoured Division and the DD swimming tanks of 27th Armoured Brigade. The 1st Special Service Brigade of four Commandos landed in the following waves to capture the strong points either side of the landing area and push inland to the Orne bridges to link with the paratroops that had captured them during the night.

Facing the landings was *Generalleutnant* Wilhelm Richter's *716. Infanteriedivision*. Although the division has been in Normandy for more than two years, its troops have spent most of their time labouring on fortifications and little time on combat training. Sword Beach fell in the sector of *736. Grenadierregiment*, in particular *Hauptmann* Heinrich Kuhtz's 10th Company.

The defenders were well dug in and well supported with machine-guns, mortars and artillery (every gun in the division was sited to fire on the beaches), but would need immediate reinforcement to prevent the enemy getting ashore. Unfortunately, the division lacked any form of motorised reserve. Worse, the closest reserve to hand was *642. Ost Bataillon*, a battalion of Russian 'volunteers' recruited from the prisoner of war camps.

The nearest armoured troops were *21. Panzerdivision* south of Caen. This division was fairly well equipped and trained, but could not be expected to reach the beach in strength until late in the afternoon.

Whatever the outcome, Sword Beach was looking to be a hard fight and the troops knew it. Major 'Banger' King commanding A Company of the East Yorks, like Wolfe at Quebec, read to his men as the landing craft approached the beach. Shakespeare's stirring speech from King Henry V echoed in their ears as they stormed ashore.

'On, on, you noblest English!
Whose blood is fet from fathers of war-proof,
Fathers that, like so many Alexanders,
Have in these parts from morn till even fought,
And sheath'd their swords for lack of argument: ...

I see you stand like greyhounds in the slips,
Straining upon the start. The game's afoot:
Follow your spirit, and upon this charge
Cry 'God for Harry, England, and Saint George!'

Refighting Queen Red

The battle for Queen Red Beach was a dramatic example of courage under fire and the value of specialised assault armour in amphibious assaults. The Hit the Beach mission on page 29 is ideally suited to refighting this battle in miniature.

Although the assault was made by a full battalion of infantry, it is easiest to scale the battle down to a company-sized action with each platoon in the game standing in for a full company in the historical battle. Obviously the battlefield has to be scaled down to match as well.

The following two pages give a table layout covering all of the important features of Queen Red Beach and strongpoints Cod and Sole (Daimler remains off the table to the east of Sole) and a list of the forces from 3rd Division and *716. Infanteriedivision* that were engaged in the fighting there.

QUEEN RED, 6 JUNE 1944

STRONGPOINT COD

The defences of Strongpoint Cod (known as *Wiederstandnest 20* or *WN 20* to the Germans) are formidable. Only the eastern part of Strongpoint Cod is shown as the rest faces Queen White beach where the 1st Battalion, The South Lancashire Regiment is landing.

The anchor of Strongpoint Cod is an 8.8cm PaK bunker sited to fire along the beach. This is supported by a 5cm KwK bunker sited to fire along the beach to each side and inland, but completely protected from attacks coming directly off the beach by an impenetrable wall that covers its beach frontage. East of this the sea wall is covered by a Tank Turret bunker mounting an old Renault R-35 turret.

Cod is protected by mines, barbed wire and trenches making it almost impervious to frontal assault.

Further inland beyond the flooded marshes, Strongpoint Sole (*WN 14*) is the company HQ position. This strongpoint is mainly a command centre and has little in the way of defences aside from barbed wire.

Strongpoint Daimler (*WN 12*) houses four artillery pieces in formidable emplacements sited to fire out to sea and also cover the landing beaches at Sword Beach. Daimler is positioned off the table just east of Sole, but can fire artillery bombardments to anywhere on the table.

British 3rd Division

2nd Bn, East Yorkshire Regt, 8th Bde

Battalion HQ
A Company
B Company
C Company
D Company
6th (Pioneer) Platoon
A Company, 2nd Bn, The Middlesex Regt
45th Battery, 20th Anti-tank Regt, RA
76th (Highland) Field Regt, RA
B Squadron 13/18th Hussars
77th Assault Squadron, RE
A Squadron, 22nd Dragoons
2 Troop, 4 Commando

Rifle Company (Confident Trained) (page 44 Turning Tide)

Company HQ
Rifle Platoon (at full strength)
Rifle Platoon (at full strength)
Rifle Platoon (at full strength)
Rifle Platoon (at full strength)
Pioneer Platoon (at full strength)
Machine-gun Platoon (at full strength with carriers)
Assault Anti-tank platoon (SP), RA (4x M10 SP guns)
Assault Field Battery, Royal Artillery (1x Gun Troop of Priest SP guns)
DD Armoured Platoon (3x Sherman DD tanks)
Breaching Group (2x AVRE Sections)
Breaching Group (3x Sherman Crab flail tanks)
Commando Company (1x Section)
This platoon must not be in the first wave.

The first wave ashore on Queen Red Beach was made up of soldiers from the 2nd Battalion, The East Yorkshire Regiment with armoured support from 13/18th Hussars, 22nd Dragoons, and 77th Assault Squadron. Following waves came from the 1st Battalion, The Suffolk Regiment supported by the machine-gunners of the 2nd Battalion, The Middlesex Regiment.

German 716. Infanteriedivision

III/736. Grenadierregiment

Bataillongruppe (Strongpoint Sole)
Granatwerferzug (Strongpoint Cod)
10. Kompanie (Strongpoint Cod)
12. Kompanie (Strongpoints Cod and Sole)
736. Panzerjägerzug (Strongpoint Cod)
4/1716. Artillerieregiment (Strongpoint Daimler)

Alarmheiten

4. Kompanie, 736. Grenadierregiment
3/642. Ost Bataillon
736. Sturmzug
3. Kompanie, 22. Panzerregiment, 21. Panzerdivision

Festungskompanie (Confident Trained) (page 22 Earth & Steel)

Company HQ (with one HMG Nest, at A)
Company HQ Mortar Section (2x 8cmGW42, attached to Festungs Grenadier Platoon, at B)
Festungs Grenadier Platoon (at full strength, at B)
Machine-gun Platoon (with three HMG Nests at B)
Festungs Anti-tank Gun Platoon (two 7.5cm PaK97/38 guns, at C)
Festungs Artillery Battery (off table, Observer teams at A & B)

Reserves

Festungs Grenadier Platoon (at full strength)
Ost Grenadier Platoon (at full strength)
Festungs Scout Platoon (one Squad with Command Panzerknacker SMG)
Panzer Platoon (four Panzer IV tanks rated Confident Veteran)
The Panzer Platoon must be the last reserve platoon to arrive.

The German defenders were from *III. Bataillon, 736. Grenadierregiment* of *716. Infanteriedivision*. All of the division's guns were deployed well forward where they could fire on the defenders as they landed. The immediately available reserves were mainly unsupported infantry, although the tanks of *21. Panzerdivision* arrived towards the end of the day.

What If Battles

While the forces shown above represent the forces that actually landed on Queen Red Beach and those that opposed them, you can also play the scenario with different forces to see how the battle would have turned out if the commanders had committed different forces to the battle.

An Allied force attacking at Queen Red Beach should consist of 1875 points based on an Infantry Company.

An Axis force should consist of 1250 points based on an Infantry Company. Any tank or self-propelled gun platoons should be the last to appear from reserve. Remember to retain all the fortifications on the map on page 34.

3

THE BATTLE FOR EASY GREEN

In the early morning hours of 6 June German soldiers of *352. Infanteriedivision* peer anxiously out to sea from *Wiederstandnest 66* (Strongpoint 66). A myriad of ships fills the horizon as far as their eyes can see. Frantic calls flood the telephone switchboards as the beach defences report the massive invasion fleet to a disbelieving headquarters. Bright flashes pick out the warships. Seconds later deafening explosions rock the ground. The invasion is happening, now!

Soldiers of the 116th Regiment, 29th Infantry Division clutch their weapons, shifting nervously, lost in their thoughts within their LCVP landing craft, the small steel boats pitching and bouncing in the rough surf. Their first combat of the war weighs heavily upon these uncertain men. With a final roar of engines, the hulls scrape on the sand as the landing craft drop their ramps into the foaming surf.

Seasick soldiers awkwardly clamber down the ramps, assault vests weighing 60lbs (30kg) strapped to their backs. They have come for battle, loaded with ammunition and explosives, flame-throwers and Bangalore Torpedoes. Trudging through the surf to the soft sand, the soldiers drop to a knee at the first obstacle belt, waiting for their leaders, surveying the bare beach. Fully a quarter of a mile (400m) away the shingle bank is the first available cover. Something doesn't look right, but given the lack of fire perhaps the Germans have fled as many have said they would.

Up on the bluffs the gunners finger the triggers of their MG42 machine-guns, waiting, watching, while the American soldiers below form their skirmish lines and the landing craft begin to withdraw. First one fires, then all at once, filling the air with bullets, scything down the *Amis* at 1200 rounds per minute. Soldiers fall here and there upon the beach and still they fire.

Leaders go down in a hail of angry lead. Explosions fill the air with saltwater and sand. Everywhere at once men scream in pain as they fall. The beach becomes a killing ground. One by one the soldiers dash or crawl to the relative safety of the shingle bank. The remaining leaders of Fox and George Companies look around and realize they are on the wrong beach. The assault wavers in the face of destruction—the men disorganized, shocked by the death and destruction littering the beach behind them.

More LCVP landing craft are coming ashore. Fresh troops spill out, running, dying, struggling through the bullet-torn hell to the shingle bank. Even there soldiers are dying as machine-guns fire along the length of the beach ripping into the bodies huddled behind the meagre cover.

In the midst of the carnage some men rise to the occasion. Major Sidney Bingham, commanding officer of 2nd Battalion, finds himself a squad leader. With a small group of men he storms a machine-gun nest in the basement of a three-story house, opening the way forward. Four Sherman DD tanks of the 743rd Tank Battalion lend a hand. One by one, they silence the machine-gun bunkers. Slowly, the American infantry climb the steep bluffs, moving through the thin crust of defences and into the French countryside.

HIT THE BEACH MISSION

The Easy Green scenario uses the Hit the Beach mission on page 29.

BEACH DEFENCES

The defences of *Wiederstandnest* 66 (Strongpoint 66) are still under construction. Even so, much of the work is done, providing the soldiers with a very defensible position.

The completed defences consist of one pedestal mounted 5cm KwK in an open bunker just off the beach. Three heavy machine-guns in open Tobruk-style bunkers provide enfilade fire support for the 5cm KwK and the obstacle belt. In the actual strong point a *Panzerstellung* tank-turret provides further firepower. Communication trenches ring the entire strongpoint behind strong barbed wire obstacles.

At the top of the draw is another defensive network of trenches and wire. These defences protect the 28cm sWG41 rocket launchers aimed at the beach.

LES MOULINS OR D3 DRAW

The gully known to the Americans as the D3 Draw is the only way off the beach for motorized traffic. The steep bluff behind the beach stops all vehicle movement up its' face.

Realizing the importance of the draw and its road the Germans blocked the way up it with minefields and anti-tank obstacles effectively making it impassable without considerable engineering work.

The bluffs are Very Difficult Going and are Impassable to all vehicles.

GERMAN FORCES AT EASY GREEN

Background

In May 1944, the 5th Company, *916. Grenadierregiment*, relieved a platoon of the second-rate 10th Company of *726. Grenadierregiment* manning Strongpoint 66 east of Les Moulins draw (overlooking the beach known as Easy Green to the US Army.) When the battle started the quality of these troops would be the biggest surprise inflicted on the Americans.

Despite Rommel's orders, only one platoon actually occupies the beach defences, the rest of the company making up a reserve to the rear, ready to support both Strongpoint 66 and Strongpoint 68 (on the opposite side of the draw) as the situation dictates.

Further to the rear, in Strongpoint 67, a static rocket launcher battery provides heavy and immediate fire support.

Force Tactics

While your defences look sparse at first glance, they provide a rock-hard defence. The bunkers and fortifications deny the enemy troops the easy way off the beach by going up the road. Instead the enemy will have to advance slowly through your defences, each turn fighting to gain a few metres of terrain.

Your advantage is that the enemy has to continually advance, and in so doing loses all cover and concealment. To win, all you need to do is keep your opponent pinned down on the beach, preventing him from moving forward. The static rocket launchers are excellent for this.

Once this is accomplished you will have time to get your reserves up and into position to deny him the objectives. Your men have prepared their positions, trained hard, and are now ready to fight.

German 352. Infanteriedivision

Wiederstandnest 66	Festungskompanie (Confident Veteran) (page 22 of Earth & Steel)
5. Kompanie, 916. Grenadierregiment	Company HQ (with two 8cm GW42 mortars, HQ trained as Panzerknacker SMG teams, at A)
1. Zug, 5. Kompanie	Veteran Festungs Grenadier Platoon (at full Strength, with MG teams, HQ trained as Panzerknacker SMG teams, at A)
Befestigungen	**Fortifications**
VF600 casement	5cm KwK Nest
Panzerstellung	APX Tank-turret bunker
Tobruk Stellungen	Three HMG Nests
84. Nebelwerfer Abteilung	Four 28cm sWG41 Nests
Verstärkungen	**Reserves**
3. Zug, 5. Kompanie, 916. Grenadierregiment	Veteran Festungs Grenadier Platoon (at full Strength, with MG teams, HQ trained as Panzerknacker SMG teams)

Much to the surprise of the 116th Infantry Regiment, the German defenders at *Wiederstandnest 66* are men from 5. *Kompanie,* 916. *Grenadier* Regiment of 352. *Infanteriedivision*. This newly formed division is far better equipped and trained than the static coastal divisions they expected to find. Although the defences are only manned by a single platoon, of infantry, they are led by veterans of the Eastern Front with plenty of machine-guns and ammunition. Without a doubt, *Wiederstandnest* 66 will cause heavy losses to the troops landing in front of it.

What If Battles

Flames Of War gives you an excellent forum to test out not only what happened in the battle for Easy Green, but what might have happened as well.

An example of this is to take the German reserves and drop the reserve infantry platoon, replacing it with two StuG assault guns, to represent *Kampfgruppe Meyer* coming to the beach instead of wandering in the countryside for the whole battle.

You can play the scenario with other forces entirely as well. If you do so, the attacking force consists of 1500 points while the defending force should consist of 1200 points. Any motorised, self-propelled gun, or tank platoon should be the last to arrive for the defending force. Remember to retain all the fortifications on the map on page 37.

US FORCES AT EASY GREEN

BACKGROUND

Major Bingham's 2nd Battalion 116th Infantry Regiment was supposed to land in the Dog Red sector of Omaha Beach just after dawn on D-Day, 6 June 1944. Plans have a way of not working out once the battle begins.

The trouble started early. A strong current was running up the Channel from the south, pushing the landing craft off course. Fox and George Companies landed in the Easy Green sector, a more heavily defended portion of the beach.

The trouble continued as heavy seas made it unsafe to launch the Sherman DD tanks to swim ashore ahead of the infantry. Instead they landed directly onto the beach from their landing craft—in hindsight the correct decision, but one that would leave the infantry unsupported for the first critical minutes ashore.

Wading ashore, the troops found themselves on an unfamiliar beach under heavy fire. Soon the tanks would join them, only to be stopped by the combination of the anti-tank obstacles and the high bluffs.

Forced to stay on the beach, the tanks provided what fire support they could as the infantry advanced up the bluffs, attacking the strongpoints from the rear.

FORCE TACTICS

Easy Green sector is well defended. You will have to use each of your units' capabilities to their best if you are going to win. But always remember, you are there to attack, and attack again. A timid nature will get you nowhere.

Each boat section contains a great deal of capability packed in a small force. They have weapons for every eventuality, along with specialised equipment like Bangalore Torpedoes to blow gaps in barbed wire and demolitions to destroy bunkers. These men are your assault troops. Get them in close as fast as possible and use their flame-throwers and demolitions to take apart the enemy defences. Don't forget to use your light mortars and machine-guns to pin the enemy first.

The Sherman DD tanks give you the capability to keep bunkers pinned down until the infantry can close with and destroy the bunkers. Then they climb off the beach and move up the road. If the bunkers are still fighting when your anti-aircraft artillery half-tracks land, they can be very effective in pinning down bunkers as well.

Finally, don't forget your naval fire support. A well-placed barrage can turn the tide of battle.

US 29TH INFANTRY DIVISION

II/116TH RIFLE BATTALION	ASSAULT COMPANY (CONFIDENT TRAINED) (PAGE 172 TURNING TIDE)
Battalion HQ	Company HQ
Fox Company, 2 Boats	2 Boat Sections (at Full Strength)
George Company, 4 Boats	4 Boat Sections (at Full Strength)
2nd Platoon, Able Company, 743rd Tank Battalion	DD Tank Platoon (with four M4 Sherman DD tanks)
2nd Platoon, Charlie Company, 112th Engineer Combat Battalion	Assault Engineer Combat Platoon (at Full Strength with four Bazooka teams and M4 Sherman dozer)
Able Battery, 111th Field Artillery Battery	Field Artillery battery (at full strength with DUKW trucks)
1st Platoon, Baker Battery, 467th Anti-aircraft (AW) SP Battalion	Corps Anti-aircraft Artillery (Self-propelled) Platoon (with two M16 MGMC and two M15 CGMC half-tracks)
USS Carmick	Naval Gunfire Support (see page 25)

The first wave ashore is not Easy Company of the 116th Infantry Regiment as planned. Instead, due to the strong crosscurrent, the first troops ashore on Easy Green are elements of Fox and George Companies. As the DD tanks disembark from their landing craft directly on to the shore, having chosen not to attempt to swim ashore in the heavy seas, they bring the defences under fire. With their support, the infantry wade ashore and fight their way off the beach.

4

POINTE DU HOC

When the Allies planned the Normandy landings, they knew they would have to destroy the German gun batteries at Merville in the British sector and Pointe du Hoc in the US sector. Both were capable of wreaking serious damage on the invasion fleet.

Thus an epic tale of luck and tenacity was written. For Pointe du Hoc, the story was one of bravery, courage, and a great deal of luck. Without any one of these the mission of 2nd Ranger Battalion to destroy the cliff-top guns might have turned out very differently.

The 2nd Rangers planned to send their Dog, Easy, and Fox Companies in as the first wave. Charlie Company would go in with the 29th Infantry Division to destroy another battery. This left Able and Baker Companies in reserve. These companies would reinforce the rangers if signalled to do so, otherwise they would reinforce the landings at Omaha. Disregarding the direct orders of his superior officer, Lieutenant Colonel Rudder accompanied his men in their assault.

Prior to the landing, the *USS Texas* brought its 14-inch guns to bear on the defences, its huge shells carving out a moonscape of giant craters. Promptly at 0630hrs the 'Old Battleship' shifted its fire to new targets to allow the rangers to land. Unfortunately though, the rangers' landing craft had run off course and were struggling against the current to make their way back along the coast.

The German defenders were not idle. With the bombardment over long before the rangers arrived, they reoccupied their positions. The rangers finally arrived at the beaches below the cliffs half an hour late. The reserve was already on its way to Omaha Beach. Firing machine-guns and lobbing grenades down the cliff, the defenders struck at the rangers as they landed. Despite the heavy fire, the rangers began firing rockets trailing ropes up the cliffs and began climbing.

After reaching the cliff-top, small groups of rangers set about carrying out their mission without waiting for the soldiers struggling up the ladders behind them. Hopping from crater to crater, they advanced under heavy German fire. Reaching their objectives, they found that the guns were no longer there! Weeks before, the Germans had repositioned the guns to save them from the Allies' incessant aerial bombardments.

Colonel Rudder then sent his men out to find the guns. Following wheel tracks in the soft mud, a pair of rangers came upon the guns in positions further inland, ready to fire. Working frantically and with the explosives they had with them, they disabled a single gun. Running back to their unit they grabbed all the thermite grenades available and returned to the guns. From the other edge of the clearing German voices drifted over to the two men. Quickly they placed the charges destroying the remaining guns.

At almost the same time a large explosion rocked the ground and a tremendous fireball filled the sky—another team of rangers had discovered the ammunition dump and finished it off. With the mission complete the rangers awaited the inevitable counterattacks and their relief.

German reinforcements from *II Bataillon 914. Grenadier-regiment* descended on the ranger's positions. Over and over the ever thinning ranks of rangers beat them off. Soon the poorly-motivated *439. Ost Bataillon* joined the fray. The rangers withdrew behind the fortifications of the base perimeter, prepared to make a final stand. After two days of relentless assaults, the relieving force finally arrived.

Despite bad luck leading them astray and landing late only to find the guns moved, the rangers had succeeded through desperate bravery and courage. The cost was high. By the time relief arrived two days later, of the more than 200 rangers that landed, only 50 were still capable of combat.

POINTE DU HOC MAP

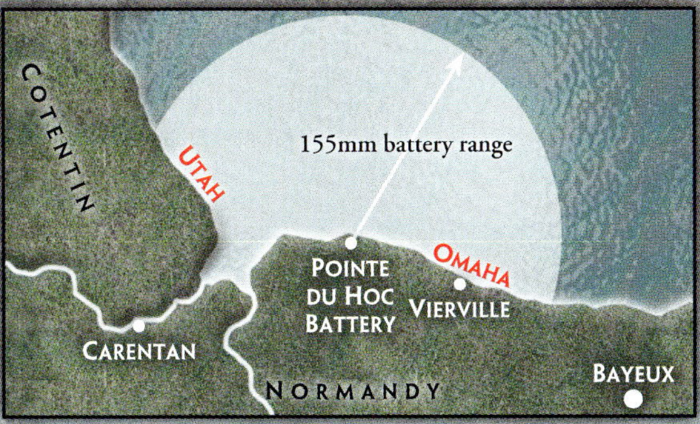

Pointe du Hoc provided the German gunners with an easily defended vantage point with which to engage any Allied fleet, making it a prime target on D-Day.

LCA (Landing Craft, Assault) Roster

LCA668 D Company	LCA861 E Company	LCA883 F Company
LCA858 D Company	LCA862 E Company	LCA884 F Company
LCA722 E Company	LCA888 E Company	LCA887 F Company

LCA860 D Company, became swamped and sunk before landing.

POINTE DU HOC TABLE

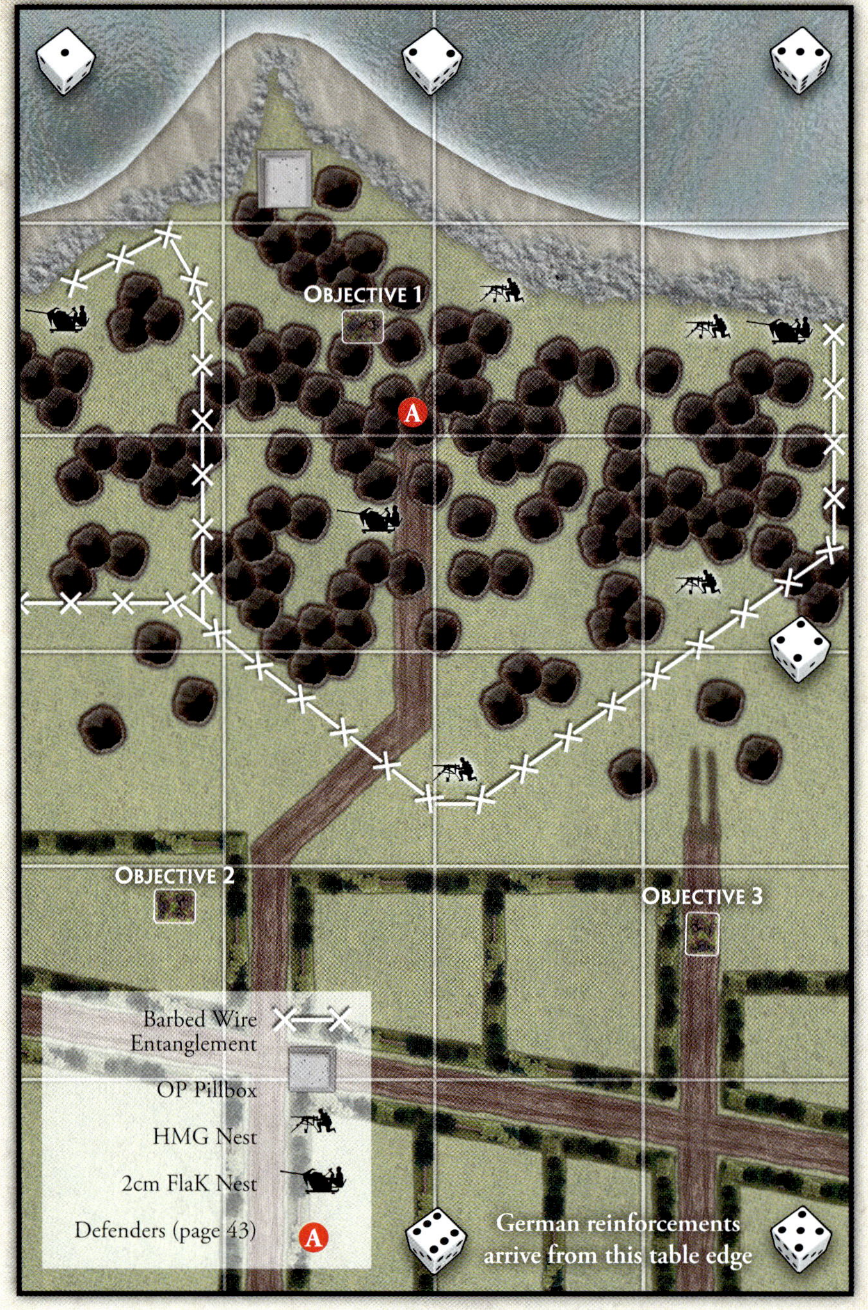

The defences of the Pointe Du Hoc battery use the terrain well. Situated on a point jutting into the English Channel atop sheer cliffs, the battery appeared impervious to attack from the sea. A series of barbed wire obstacles protect against attack from its landward side.

The defences are held by a platoon of infantry with two heavy machine-guns covering the sea approach and two more facing landward. Three 2cm anti-aircraft guns in concrete pits have recently been added to deal with the constant Allied air attacks.

Cliffs

The cliff is sheer and tall making communication up and down impossible.

Teams on the beach may not assault or be assaulted, nor shoot or be shot at. Teams at the top of the cliff are never within Command Distance of those at the bottom.

Shell Craters

Previously trenches linked the bunkers and shelters. Now the entire area looks like a moonscape.

The entire area between the cliffs and the hedgerows is covered in shell craters. These give teams Concealment. Teams in shell craters also gain Bulletproof Cover except from Defensive Fire.

OP Bunker

The OP bunker on the point would have controlled the guns had they fired, but played no part in the battle.

The OP Bunker is Impassable and cannot be occupied.

THE BATTLE FOR POINTE DU HOC

The Pointe du Hoc scenario recreates this battle. With so many things that could have and did go wrong for the rangers, and a few that went right too, this scenario has a large element of chance in it, much like the real thing. Luck will play a large part setting the scene for the battle, but it is your skill as a commander that will determine the outcome.

The Pointe du Hoc scenario uses the **Colonel Rudder**, **Don't Wait for Me**, **Random Objective**, **Ranger Reserves**, **Protect the Guns** (see page 44), **Preliminary Bombardment** (see page 26), **Delayed Reserves**, and **Prepared Positions** (see pages 264 to 269 of the rulebook) special rules.

YOUR ORDERS

US

You must climb the cliffs of Pointe du Hoc to find and destroy the guns before they can bombard the invasion fleet. Living to tell the tale would be a nice bonus!

GERMAN

You must stop the rangers from destroying the guns and then wipe them out for daring to invade Fortress Europe.

PREPARING FOR BATTLE

1. Set up the table as shown on page 42, including placing the Random Objective.

2. The German player deploys their garrison platoon (see below) inside the barbed wire perimeter. The remaining forces are held off table in Delayed Reserve.

3. The US player now places Colonel Rudder and the landing force on the beach against the cliff. The Reserve Force will arrive during the game using the Ranger Reserves special rule.

The Rangers will climb the cliff using their Mountaineers special rule (see page 61 of the rulebook).

BEGINNING THE BATTLE

1. The German player starts the game in prepared positions, so their troops are Entrenched in foxholes, trenches and craters, and are Gone to Ground.

2. The US player conducts a Preliminary Bombardment on the defending German forces.

3. The US player takes the first turn.

ENDING THE BATTLE

The battle ends when either:

- The US force has been destroyed or
- On the beginning of the US player's 13th turn.

DECIDING WHO WON

Due to the chaotic nature of the fight and the variable objectives and reinforcements, it is impossible to clearly define victory or defeat for either side.

The US rangers' goal is to destroy the guns and limit their losses so they may fight another day. If things go well, destroying the guns should be easy, and keeping the whole force alive until relieved will be the main goal. If the guns turn out to be hard to find, simply destroying them with the last surviving rangers will be a major achievement.

The German defenders must protect the guns at all costs, or failing that, destroy the invaders, or at least inflict significant casualties on them. Depending on where the guns are, the first may prove impossible, making revenge and killing as many rangers as possible the order of the day.

BOCAGE HEDGEROWS

The hedgerows at the south end of the table are bocage hedgerows. These are taller than a tank, Very Difficult Going, and provide Bulletproof Cover to troops sheltering behind them. Teams must start their movement adjacent to a bocage hedgerow to cross it. The full Bocage rules are provided in more detail in the rulebook.

GERMAN 352. INFANTERIEDIVISION

SICHERHEITSZUG	**FESTUNGSKOMPANIE (SEE PAGE 22 EARTH & STEEL)**
2. Zug, 1. Kompanie, 726. Grenadierregiment	Veteran Festungs Grenadier Platoon (at full strength, rated as Confident Trained, at A)
VERSTÄRKUNGEN	**RESERVES**
1. & 3. Zug, 3. Kompanie, 914. Grenadierregiment	2 Veteran Festungs Grenadier Platoons (at full strength, with MG teams, HQ equipped as Panzerknacker teams, rated as Confident Veteran)
1. Zug, 4. Kompanie, 914. Grenadierregiment	Festungs Machine-gun Platoon (with 2 Machine-gun Sections, rated as Confident Veteran)
2. Zug, 1. Kompanie, 439. Ost Bataillon	Ost Grenadier Platoon (at full strength, rated as Reluctant Trained)

POINTE DU HOC SPECIAL RULES

Colonel Rudder

Lieutenant Colonel Rudder led his men ashore on D-Day where he discovered that all of the radios operating on navy frequencies were dead. However, one of his men had brought along an old signalling lamp. Through him, Rudder controlled the fire of the *USS Satterlee* by Morse Code.

Colonel Rudder is a Warrior Higher Command Carbine team and the Observer team for the Naval Gunfire Support.

Don't Wait For Me

The Pointe du Hoc raid was carefully planned. Every ranger knew what needed to be done and had memorised the photographs and sand tables showing their objectives. The first troops up the cliff lead the way, regardless of rank.

When a platoon climbs the cliff using the Mountaineers special rule, the first team up the cliff will always be the Platoon Command team. If the Platoon Command team does not pass its Skill Test to climb the cliff, swap the Platoon Command team with a team that did climb the cliff. In addition, Ranger Platoon Command teams do not need to keep half of their platoon In Command when they move. They can move off with whatever force they have available.

Random Objective

The rangers were surprised to find the gun positions empty and had to search for the guns. Likewise the German reserves had no idea that they had been moved.

At the start of the game neither player knows which of the Objectives on the table is actually the guns. Before the game take two pieces of paper. Mark one of them and then randomly place a piece of paper under each of Objectives 2 and 3. The marked Objective represents the guns. Neither player should know which is the marked Objective.

The rangers must take Objective 1 before they can take Objectives 2 or 3. When the rangers take Objective 2 or 3, they reveal the piece of paper under it. If it is the marked paper, they have found the guns and destroyed them. The German player is not allowed to look at the piece of paper until the Ranger player reveals it by taking the Objective (having already taken Objective 1).

Ranger Reserves

Rangers from everywhere tried to get to the battle. Some arrived by boat, others by land after a difficult journey.

Rangers follow the normal Delayed Reserves and Scattered Reserves special rules with the following changes:

The Ranger player only rolls one die to receive Reserves each turn, rather than increasing the number of dice each turn, and requires a roll of 6 to receive Reserves, instead of the usual 5+.

When reserves arrive roll another die, consult the map and place the reserves on that appropriate table edge or corner based on the die roll. If reserves arrive from a corner, they must enter the table within 16"/40cm of the corner.

The Tank Platoon must be the last to arrive from Reserve. In addition, if the Scattered Reserve roll to find out where it arrives is a 1, 2, or 3, they do not arrive this turn. Instead, they roll again for their arrival point next turn.

Protect the Guns

The German defenders have lost all communications and are fighting in a moonscape out of touch with friendly troops. Unaware of what is happening elsewhere on the battlefield, they will fight until they themselves are defeated.

There is no German Company Command team in this scenario. However, the German force will always pass any Company Morale Check it is required to take.

US 2nd Ranger Battalion

2nd Ranger Battalion	Ranger Companies (Fearless Trained) (See page 186 of Turning Tide)
Colonel Rudder	Higher Command Carbine and NGFS Observer team
Mortar Platoon	Ranger Mortar Platoon (at full strength)
Dog Company	Ranger Company (with two platoons and two bazookas)
Easy Company	Ranger Company (with two platoons and two bazookas)
Fox Company	Ranger Company (with two platoons and two bazookas)
USS Satterlee	Naval Gunfire Support (Colonel Rudder is the Observer team)
Reserve Force	**Reserves**
Able Company	Ranger Company (with two platoons and two bazookas)
Baker Company	Ranger Company (with two platoons and two bazookas)
Charlie Company	Ranger Company (with two platoons and two bazookas)
3rd Platoon, Able Co., 743rd Tank Battalion	DD Tank Platoon (with 3 M4A1 Sherman DD tanks)

2ND & 5TH RANGER BATTALIONS

The need for specially-trained troops to conduct dangerous missions in difficult terrain was not lost upon the US Army. Learning from their allies and enemies alike the Army began to create new elite ranger battalions in Scotland in early 1942. As the fighting was anticipated to shift from the Mediterranean to northern France, the need for more ranger units also grew. Thus the 2nd and 5th Ranger Battalions were formed in April and September 1943 at Camp Forrest, Tennessee. These two battalions embarked on the now legendary ranger training program to get them ready for combat.

With their training complete, they headed for England. Their arrival in September of 1943 was just in time to begin immediate preparation for Operation Overlord, the invasion of France. The higher command had the perfect target for the rangers. They would destroy the coastal gun battery at Pointe du Hoc.

Come D-Day it would be 2nd Ranger Battalion that became famous for their exploits against this gun battery. There in the early morning hours Dog, Easy, and Fox Companies scaled the cliffs, and against overwhelming odds fought their way through the defences to destroy the guns, despite them having been moved from their positions to a location away from the strongpoint. By doing so they become the first United States unit to fulfil its D-Day mission. Individual courage, guts and determination allowed the rangers to accomplish the task at hand, a testament to their training.

Elsewhere, Able, Baker and Charlie Companies of 2nd Ranger Battalion and the 5th Ranger Battalion landed with the 29th Infantry Division at Omaha beach. Pinned down by the horrendous fire coming from the bluffs above the seawall the Rangers waited with the rest of the survivors of the 116th Infantry Regiment. Their wait would not be long.

Walking up the beach with a purpose, General Cota surveyed the soldiers huddled against the sea wall and still standing addressed them as hot lead flew all around. Lieutenant Colonel Max Schneider of the 5th Rangers stood up to receive his orders. Only later would he explain why he did so, saying 'You can't very well take orders sitting down.' General Cota asked which unit this was. On being told 5th Rangers, he said, 'Well then rangers, lead the way'.

With General Cota's words still in their heads the rangers scrambled over the seawall and through the wire obstacles to scale the bluffs. They led the way into the French countryside and all the way to the town of Vierville, helping to take the draw that would allow vehicular traffic to begin to flow into France from the beaches. This effort would save Omaha Beach, and allow the Allies to begin the buildup that would make Operation Overlord successful.

To this day the motto of the Rangers remains the famous words of General Cota, 'Rangers, lead the way.'

RANGER BATTALION

- **A COMPANY** (3x Light or medium mortars, 2x Bazookas)
  - 1ST PLATOON (30x Rangers)
  - 2ND PLATOON (30x Rangers)
- **B COMPANY** (3x Light or medium mortars, 2x Bazookas)
  - 1ST PLATOON (30x Rangers)
  - 2ND PLATOON (30x Rangers)
- **C COMPANY** (3x Light or medium mortars, 2x Bazookas)
  - 1ST PLATOON (30x Rangers)
  - 2ND PLATOON (30x Rangers)
- **D COMPANY** (3x Light or medium mortars, 2x Bazookas)
  - 1ST PLATOON (30x Rangers)
  - 2ND PLATOON (30x Rangers)
- **E COMPANY** (3x Light or medium mortars, 2x Bazookas)
  - 1ST PLATOON (30x Rangers)
  - 2ND PLATOON (30x Rangers)
- **F COMPANY** (3x Light or medium mortars, 2x Bazookas)
  - 1ST PLATOON (30x Rangers)
  - 2ND PLATOON (30x Rangers)

Battalion Strength: 538 soldiers in 6 companies, 18 mortars, 12 bazookas.

5

WITTMANN'S WILD RIDE

Wittmann's Wild Ride recreates the epic attack by *Obersturmführer* Michael Wittmann and his lone Tiger tank against A Company of the 1st Battalion, The Rifle Brigade on the road from Villers-Bocage to Point 213.

Unlike most *Flames Of War* battles, this scenario is rather one-sided. The British forces have no officers at all and can do little but stand where they are and fight. It is Wittmann who dictates the pace of the battle as he advances along the road towards Villers-Bocage.

This makes the scenario an ideal introduction to *Flames Of War*. The new player takes the role of Wittmann and learns the rules and tactics as they attempt to destroy the trapped British force. Be warned though, it is not as easy as it looks, after all Wittmann was an ace. You may need several games to achieve the same degree of success that he did!

The game plays quickly, making it ideal for demonstrations or as a club challenge with players taking turns and then swapping roles to see how far they can get.

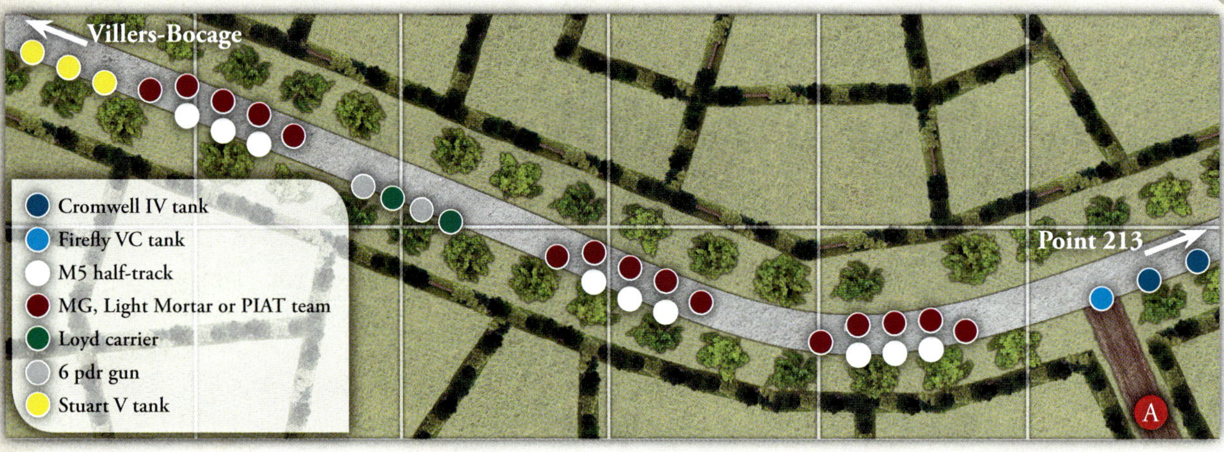

TERRAIN

The road between Villers-Bocage and Point 213 is bordered by high Bocage hedgerows. For the purposes of this scenario they are Impassable terrain and cannot be seen through. All teams must remain on the road, which is wide enough for Wittmann to pass the British vehicles. The road is paved, making it impossible for the British to entrench in the short time available.

The Wittmann's Wild Ride scenario uses the **Tea Time,** and **The Commander is Away,** special rules on page 50.

YOUR ORDERS

WITTMANN

You must attack and destroy the British column, or at the very least advance quickly to its end. Anything else dooms you to failure and results in the encirclement of the Panzer Lehr and 12 SS Divisions.

A COMPANY, 1ST THE RIFLE BRIGADE

You must destroy Wittmann's Tiger and keep the road open to A Squadron 4 CLY so that the advance to Caen may resume.

PREPARING FOR BATTLE

1. Set up the table as shown above. It is set up on a 2'/60cm by 6'/180cm table, each square is 12"/30cm on a side.

2. The British player now sets up their force as shown. The details of their force are given on page 50.

3. The German player places Wittmann's Tiger tank on the road at position A on the map above.

BEGINNING THE BATTLE

1. All British vehicles start the game Bailed Out and all British platoons start the game Pinned Down in accordance with the Tea Time special rule on page 8.

2. Under The Commander is Away special rule, the British have no officers, so will not move during the game, except in assaults.

3. The German player takes the first turn. He counts as if he moved in his first Movement Step, even if he did not, as Wittmann drives onto the battlefield.

ENDING THE BATTLE

The battle ends after both players have had 10 turns.

DECIDING WHO WON

The British player wins if they can destroy Wittmann, thus keeping the way open to the forces on Point 213.

The German player wins if they can entirely destroy the British force, saving the German flank.

If the game makes it to the end of turn 10 the result is a draw. The British still hold Villers-Bocage, but the way to Hill 213 is still blocked by Wittmann.

OBERSTURMFÜHRER MICHAEL WITTMANN

Born on 22 April 1914 in Bavaria, Michael Wittmann became perhaps the best recognised, even legendary, tank ace of World War II. His military career started in 1934 when he joined the army as an infantry soldier. This early training sparked Wittmann's interest in tanks.

In 1936 at the end of his enlistment, he joined the *Allgemeine-SS* and began his career in the *SS-Verfügungstruppe*, the military branch of the SS. When the war began, Wittmann was still in training as an armoured car commander. After only a short combat stint in Poland commanding an armoured car with just a 2cm gun, he found himself transferred to a *Sturmgeschütz* assault gun with a 7.5cm gun. While his performance in the subsequent Balkan campaign was good, it would be in Russia that his talents became obvious.

In his first summer in Russia, Wittmann was awarded the Iron Cross Second Class, soon followed by the Iron Cross First Class. In the spring of 1943, he joined the *Liebstandarte SS Adolf Hitler* Division's 13th *Panzer* Company as a Tiger tank commander. By 1944, he had the Knights Cross with Oak Leaves for his gallant service in combat.

Villers-Bocage would be his heyday. In a single day he stopped, almost single-handedly, an entire British Armoured Division and saved the flank of the Panzer Lehr Division from being trapped and destroyed, earning the Swords for his Knights Cross in the process.

He died a soldier's death on 8 August at Gaumesnil, south of Caen, leading his Tiger tanks into battle one last time.

GERMAN 2. KOMPANIE, 101. SCHWERE SS-PANZERABTEILUNG

Kompanieführer	Obersturmführer Michael Wittmann

See page 139 of *Earth & Steel* for Obersturmführer Michael Wittmann's special rules

Watching the British vehicles roll by his position, one after another, *Obersturmführer* Michael Wittmann knows he must act. His orders are to hold the hill and the town of Villers-Bocage, but the British have got there first. If he cannot immediately dislodge them then the entire eastern flank of the German defences in Normandy will be in jeopardy.

Maintaining radio silence is critical to ensure surprise for his counterattack—the British will be listening for German transmissions. Wittmann sends a runner back to von Westernhagen at battalion and calls over his tank commanders. Four Tiger tanks will deal with the British on Point 213, while he alone will counterattack towards Villers-Bocage. He will take the more difficult job as he has many times before, always leading from the front and setting the example for his men. With his own tank broken down on the march, Wittmann takes over tank 212 for the battle.

Wittmann is not concerned with the odds—he has surprise on his side. It will be tricky, but he is sure of victory.

BRITISH 7TH ARMOURED DIVISION

Veterans of many battles fought in the dry heat of the desert, the men slap each other on the back. They are A Company of the 1st Battalion, The Rifle Brigade under the command of Major J Wright. They are a force to be reckoned with, one that has seen combat and proven its mettle time after time. They know there is nothing that they cannot deal with.

Just ahead is one of the troops of A Squadron, 4th County of London Yeomanry under the command of Major P M R Scott. Their tanks should keep the enemy at bay. Now it's time for the breakfast they missed with an early start, beating Jerry to the punch. It'd be a laugh to see Rommel's face when he finds himself out-flanked and trapped!

British 7th Armoured Division

A Company, 1st Bn, The Rifle Brigade	Motor Company (Reluctant Veteran) (page 102 Turning Tide)
1 (Motor) Platoon	Motor Platoon (without Command team and its half-track)
2 (Motor) Platoon	Motor Platoon (without Command team and its half-track)
3 (Motor) Platoon	Motor Platoon (without Command team and its half-track)
14 (Anti-tank) Platoon	Motor Anti-tank Platoon (with two 6 pdr guns with Loyd Carriers, but without Command team and its Troop Carrier)
4 Troop, A Squadron, 4th County of London Yeomanry	Armoured Platoon (with two Cromwell IV and one Firefly VC tank, but without the Command Cromwell IV tank)
Recce Troop, 4th County of London Yeomanry	Recce Patrol (with three Stuart V tanks none of which is a Command team)

The British force in this mission has no Company HQ and no platoon leaders (see The Commander is Away special rule below). The organisation of the British force is covered on page 102 of *Turning Tide.*

SCENARIO SPECIAL RULES

The Commander is Away

A Squadron, 4 CLY has arrived at Point 213. With his regiment making good time and poised to strike towards Caen, Lord Cranley has called an Orders ('O') Group on Point 213, summoning all of his officers to explain the plan for the next stage of the advance.

The British force has no Company Command or 2iC Command teams, nor any Platoon Command teams.

Without a Company Command team, the British force cannot pass a Company Morale Check if it is required to make one. The entire British force will be Destroyed at the start of any British turn in which four or more of their platoons are Destroyed.

Without Platoon Command teams, the British platoons cannot move except to Launch an Assault or Counterattack (using the Carry On Sergeant special rule on page 246 of the rulebook) or Break Off or Consolidate after an assault.

Despite having no Platoon Command teams, the British platoons can still make any Motivation tests that they are required to make, as usual.

Tea Time

When the 'O' Group was called, all of the officers of A Company of The Rifle Brigade set off in their half-tracks to attend.

Meanwhile, the men settled down to make the breakfast they had missed with an early start. Gathering around small fires the men begin a brew up. It's time for tea.

All British vehicles start the game Bailed Out and all British platoons start the game Pinned Down.

Remember, platoons can Rally from being Pinned Down and Remount their vehicles without a Command team, so this penalty is only temporary.

VILLERS-BOCAGE CAMPAIGN

Wittmann's Wild Ride can be played as part of a campaign with the Point 213 and Villers-Bocage scenarios in this book. The winner of the final Villers-Bocage scenario wins the campaign. However, your chances of winning that battle will be greatly enhanced or reduced by your performance in this battle and in the Battle for Point 213 scenario on page 52.

If you decide to combine the scenarios into a campaign, then whoever wins this scenario will have extra forces available for the Battle for Point 213 scenario.

OTHER PLACES, OTHER TIMES

As it happened, the British 7th Armoured Division made the dash to Villers-Bocage. What would have happened if it had been the US 2nd Armored Division instead?

You could refight the scenario with US Armored Rifle Platoons in place of the British Motor Platoons, a US Anti-tank Platoon in place of the British one, a Tank Platoon instead of an Armoured Platoon, and a Light Tank Platoon in place of the Recce Patrol. Assume that the officers are on Point 213 receiving orders and the soldiers are making breakfast, so the special rules still apply.

Another similar battle took place on the Eastern Front when Otto Carius, another Tiger ace found a Soviet tank force settling down for the night after breaking through the German lines.

You could refight the scenario with a Tank Rider Company lining the road while their officers go forward for their orders. They could have a pair of guns from a Heavy Anti-tank Company and some T-34 tanks supporting them.

6

POINT 213

The British

Moving quickly out of the town of Villers-Bocage, A Squadron 4 CLY, veterans of the famous Eighth Army, advance down the road towards Caen. In the desert, you can see the enemy from a distance. Here the troops scan the surrounding hedges for any sign of the enemy, yet nothing meets their gaze. Finally, after days of ambushes and frustration, it's back to the best days of the desert—fast free-wheeling advances.

They have encountered no enemy all day. Villagers flock to wave at the column of troops as they pass, giving the march a festive air. 'Perhaps the road is open all the way to Caen', the desert veterans reason, an easy victory within their grasp.

Despite the general cheer, Major Scott, commanding the squadron, is feeling pushed and claustrophobic in the dense terrain. The advance is moving too fast. The recce troop sits in the middle of the column rather than scouting out the flanks and securing the route of march. Yesterday's ambush proved that the Stuarts don't have the firepower for the job.

The decision to preserve the strength of the recce troop is one that will haunt the Major as events unfold.

The column pauses at the top of the hill providing a view of the entire plain almost all the way to the coast. From the top, the city of Caen is visible in the distance. The heavy bocage breaks up some of the view, but it is spectacular none the less. Lieutenant Colonel Lord Cranley, the regimental commander, calls his officers forward to give them the next course of action. But that action will be decided for them. In the bocage another man has already begun to act in the decisive manner that has covered his gun barrel with kill rings.

The Germans

Michael Wittmann, a veteran of four years of almost constant combat, watches as the British column roars by his well-hidden tanks. With so much at stake, and his orders clearly defining his role, he decides to act immediately. A simple plan of action emerges.

His once mighty company of 14 Tiger I E heavy tanks is now reduced to a mere five as the rest straggle in from the long march. Yet Wittmann knows, even in such small numbers, a Tiger in the hands of his elite crews is more than a match for anyone. He orders four of his five tanks to take Point 213. He will take the fifth tank alone on the most dangerous part of the job, preventing British reinforcements from getting to the hill until his men get the job done. With the enemy tanks on the hill destroyed, together they can move to clear the town as well.

The four mighty Tiger tanks run along a small dirt track parallel to the main paved road the British travel on. They must take the hill and secure it against all enemy attacks. Each tank crew knows its duty to the Fatherland, each ready to sacrifice all if they must.

Two of the tanks are commanded by platoon leaders, untried in armoured combat. The other two Tigers are crewed by veterans, ready and able to do battle. Hardened steel creeps forward, ready to take out the awaiting British on the hill.

The tanks move closer along the small track. There, hidden in the bocage, sits the British advance guard, the officers in conference, the tank crews unprepared. The game of cat and mouse among the hedgerows begins.

GERMAN 101. SCHWERE SS-PANZERABTEILUNG

Obersturmführer Wittmann had only five tanks to stop the British at Villers-Bocage. He sent four (211, 221, 223, and 234) to Point 213 and took his own tank back to Villers-Bocage to stop reinforcements arriving.

TIGER 211

Tiger 211 was commanded by *Obersturmführer* Jürgen Wessel, a veteran of infantry combat on the East Front, but with no tank combat experience.

TIGER 234

Unterscharführer Herbert Steif was another hardened combat veteran of the Eastern Front. Steif had served in Tiger tanks with Wittmann earlier in the year.

TIGER 221

Untersturmführer, George Hantusch in tank 221 led *2. Zug*. After combat in Russia in 1941, Hantusch studied mechanical engineering at the SS automotive school before becoming a Tiger commander.

TIGER 223

Oberscharfürher Jürgen Brandt was a hardened combat veteran who joined the Tiger battalion to lead the reconnaissance platoon. Instead he found himself in a Tiger tank, going on to rack up an impressive 57 kills.

GERMAN 101. SCHWERE SS-PANZERABTEILUNG

2. KOMPANIE 101. S SS-PZABT	SCHWERE SS-PANZERKOMPANIE (FEARLESS VETERAN) (PAGE 140 EARTH & STEEL)
1. Zug. Tiger 211, *Obersturmführer* Jürgen Wessel	Command Tiger 1 E tank
Tiger 234, *Unterscharführer* Herbert Steif	Tiger 1 E tank
2. Zug. Tiger 221, *Untersturmführer* George Hantusch	Command Tiger 1 E tank
Tiger 223, *Obersturmführer* Jürgen Brandt	Tiger 1 E tank

The German force in this mission has no Company HQ.

POINT 213

BRITISH 4TH COUNTY OF LONDON YEOMANRY

The 4th County of London Imperial Yeomanry (King's Colonials) has a proud but fairly recent past. Originally formed in November of 1901 at Charing Cross, London, its members came from throughout the British Empire. Each squadron was drawn from overseas residents from one area. A Squadron was 'British Asian', B Squadron was 'British American' or Canadian, C Squadron was Australasian, and D Squadron was 'British African' or South African. In 1902 an E Squadron was added comprised of New Zealanders.

The regiment was renamed in 1905 as The King's Colonials, Imperial Yeomanry. While the name stuck, in 1909, each of the separate 'colony' squadrons was discontinued and the squadrons were homogenised. Then in 1910 the name was changed a final time to King Edward's Horse or The King's Overseas Dominions Regiment. After being transferred to Special Reserve in 1913, the regiment was revived for the First World War. There they fought at Mons, Loos, Ypres, Cambrai, and Italy, before being disbanded in 1924.

With the situation on the Continent heating up once again in the mid 1930's, the army was expanded and, on 27 September 1939, the 4th County of London Yeomanry (Sharpshooters) (4 CLY) were formed as a duplicate of the 3rd County of London Yeomanry (Sharpshooters) (3 CLY).

Their first battles were fought in the Western Desert with the 7th Armoured Division in Operation Crusader at the end of 1941. They continued to fight with the Desert Rats through the rest of the Desert War and into Italy before returning to Britain at the end of 1943.

There the Sharpshooters were issued fast Cromwell tanks, replacing the Sherman tanks they left in Italy. They landed in Normandy on D+1. After heavy losses at Villers-Bocage and in Operations Goodwood and Spring, they left the 7th Armoured Division and merged with 3 CLY on 1 August, 1944 to form the 3rd/4th County of London Yeomanry (Sharpshooters).

BRITISH 4TH COUNTY OF LONDON YEOMANRY

A SQUADRON, 4 CLY	ARMOURED SQUADRON (RELUCTANT VETERAN) (PAGE 84 TURNING TIDE)
Lieutenant Colonel Lord Cranley	Higher Command Daimler Dingo scout car
HQ Troop, Major Peter Scott	Company HQ (with two Cromwell IV tanks and two CS Cromwell IV tanks)
1 Troop, A Squadron	Armoured Platoon (with three Cromwell IV tanks and one Firefly VC tank)
2 Troop, A Squadron	Armoured Platoon (as 1 Troop)
3 Troop, A Squadron	Armoured Platoon (as 1 Troop)
4 Troop, A Squadron	One Independent Command Cromwell IV tank
'O' group, A Company, 1st Battalion, The Rifle Brigade	Four Independent Command MG teams with an M5 half-track each

BATTLE FOR POINT 213

With its commanding views of the surrounding terrain, the high ground is critical. By taking Point 213 you will be able to observe the ground all around, thereby dominating the road all the way to Caen.

The Battle for Point 213 scenario uses the **Meeting Engagement**, **Random Deployment** (see pages 261 and 264 of the rulebook), and **Tea Time** (below) special rules.

Your Orders

German

You must take Point 213. By holding this you give the rest of *I. SS-Panzerkorps* the opportunity to counterattack the Allied thrust, sending them reeling back from whence they came.

British

With you in possession of the hilltop, the enemy will have to withdraw, offering your division the chance to begin the pursuit of the enemy.

Preparing for Battle

1. Set up the table as shown on page 56. The table is a 4'/120cm by 4'/120cm square. Mark the table centre.

2. The British player places the objective on the table no further than 12"/30cm from the table centre.

3. The British player now deploys their entire company on the table within 16"/40cm of the centre of the table. All Independent and Platoon Command teams must be deployed within 4"/10cm of Lord Cranley.

4. The German player rolls for each of their platoons deploying them in the table corner indicated by the dice using the Random Deployment rules. On a 5 or 6 the platoon may be deployed on any corner. The Tiger tanks are set up next to or on the road at that corner on the edge of the table.

5. The German player rolls for their Tiger Ace skills.

Beginning the Battle

1. The British player starts the game Bailed Out and Pinned Down as per the Tea Time special rule.

2. The German player takes the first turn and follows the Meeting Engagement special rules.

Ending the Battle

The battle ends:

- When the Germans take the objective,
- At the end of turn 12, or
- When no German Tiger tanks remain on the table.

Deciding Who Won

The German player wins if they take the objective thereby breaking the British hold on Point 213 and freeing up the road for the advance to Villers-Bocage.

The British win if they can hold the objective thereby maintaining a jumping off point for the continued drive to Caen.

SPECIAL RULES

Tea Time

The British column having stopped for an 'O' Group has decided to have tea. The movement thus far has been uneventful and the men are hungry. Gathering around the small fires the men begin to have a cup of tea and some food.

All British vehicles start the game Bailed Out and all British platoons start the game Pinned Down.

In addition, all Independent and Platoon Command teams must deploy within 4"/10cm of Lord Cranley as they gather to receive orders.

VILLERS-BOCAGE CAMPAIGN

If you decide to play the Battle for Point 213 scenario as part of a campaign with the Wittmann's Wild Ride and Villers-Bocage scenarios, the outcome of Wittmann's Wild Ride on page 48 will change this scenario in the following ways:

- If the German player won Wittmann's Wild Ride, their force, for this scenario, is increased by a full strength Armoured SS Scout Platoon (see page 58) that will arrive using the Delayed Reserve special rule (see page 269 of the rulebook).
- If the British player won Wittmann's Wild Ride, then they may bring all the remaining forces on as reinforcements in this scenario. These reinforcements enter the table at Ⓐ any turn after the turn in which Wittmann was destroyed in the Wittmann's Wild Ride game. So, if Wittmann was destroyed on turn 6, all remaining British forces can arrive in the Battle for Point 213 game from turn 7. Of course, the arriving platoons have no Command teams, so a platoon can only arrive if an Independent team is at Ⓐ to command it. If there is no Independent team for a platoon, the platoon will wait off the table, only arriving when an Independent team is at Ⓐ ready to take command at the start of the British player's turn.
- If Wittmann's Wild Ride was a draw, then there is no change to this scenario.

Different Choices

Historically, Wittmann ordered all four of his Tiger tanks to fight at Point 213, leaving him to fight alone.

You could find out what would have happened if he had taken an additional Tiger tank with him instead. Add an extra Tiger tank to support Wittmann in the Wittmann's Wild Ride scenario to represent this option. If you do this then fight the Battle for Point 213, with one fewer Tiger tank.

Point 213

Point 213 is the top of a gently-sloping hill. The slope is so gentle that the hill had no effect on the battle.

Bocage Hedgerows

All hedgerows on the table are Bocage hedgerows. These are taller than a tank, Very Difficult Going, and provide Bulletproof Cover to troops sheltering behind them. Teams must start adjacent to a Bocage hedgerow to cross it.

The main road (shown in grey) is a major thoroughfare. Wide and paved, it offers vehicles more room to pass than most other roads in the area.

Teams on the small side tracks (shown in brown) are considered to be adjacent to the hedgerows on both sides of the road. The track is so narrow that vehicles may neither pass nor turn on them.

The full Bocage rules are provided in more detail in the rulebook.

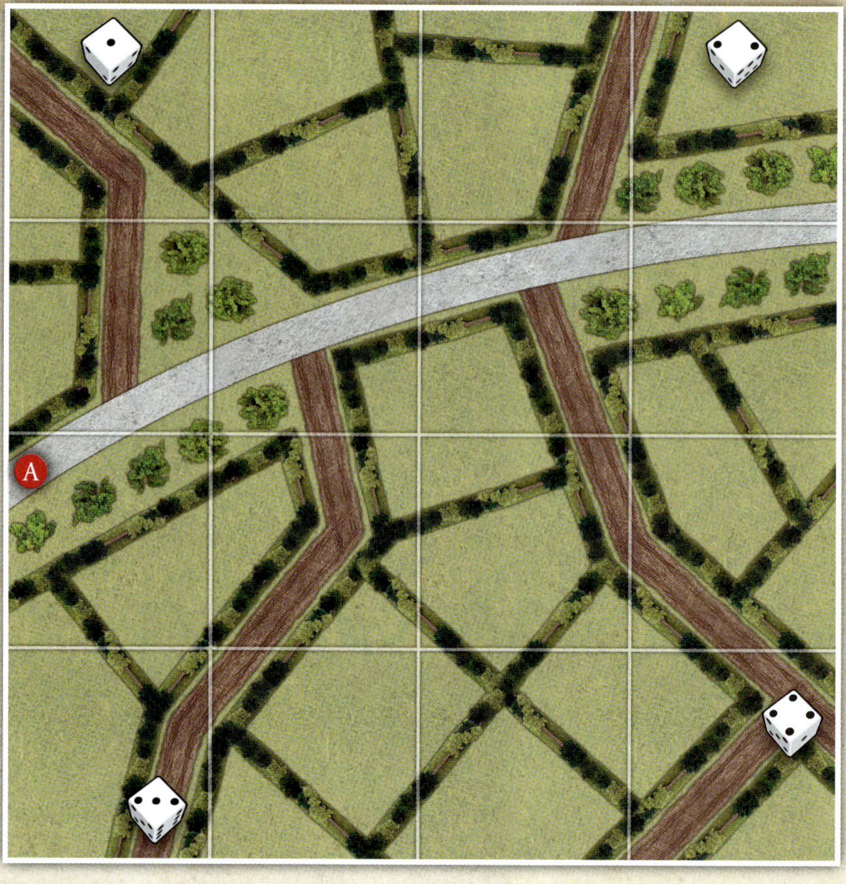

OTHER PLACES, OTHER TIMES

As with Wittmann's Wild Ride you can try this scenario with other forces to see what would have happened had someone other than the British 7th Armoured Division made the dash to Villers-Bocage.

You could refight the scenario with a US Tank Company from the 2nd Armored Division. Replace the Armoured Squadron with a Tank Company equipped entirely with M4A1 Sherman tanks armed with 75mm guns. The independent infantry teams are all Rifle teams mounted in M3 half-tracks. The soldiers are making breakfast, so the Tea Time special rule still applies.

One of Otto Carius's battles on the Eastern Front involved him and his company facing a Soviet tank force settling down for the night in a village in the middle of a forest.

You could refight the scenario with a couple of Tank Companies equipped with T-34 tanks (a few armed with 85mm guns) and a Guards Heavy Tank Company with IS-2 heavy tanks lining the village's only road while their officers gather to get their orders.

7

VILLERS-BOCAGE

PANZER LEHR ARRIVES

After his Tiger tank was damaged during his solo attack on Villers-Bocage, *Obersturmführer* Wittmann retired with his crew on foot. Realizing the gravity of the situation, Wittmann headed directly to the headquarters of the *Panzer Lehr* Division where *Generalleutnant* Fritz Bayerlein was scrounging together a force of Panzer IV tanks to take on this threat to his rear.

Wittmann immediately headed back to Point 213 in a *Schwimmwagen* amphibious jeep to brief the gathering forces of *101. Schwere SS-Panzerabteilung*. Meanwhile a *Kampfgruppe* of fifteen Panzer IV tanks and a small group of infantry under the command of *Major* Helmut Ritgen advanced towards Villers-Bocage. Their mission was to seal off the exits from the town to the west and south. The *Kampfgruppe* soon came under fire from hidden anti-tank guns and began to skirt the defences, moving southeast. Near Point 213 they linked up with the SS Tiger tanks.

THE BRITISH DIG IN

After running Wittmann out of Villers-Bocage, the British begin organising an attempt by B and C Squadrons 4 CLY to rescue A Squadron trapped on Point 213. But long before they are ready, A Squadron sends its last frantic message and then all contact is lost. With the fall of Point 213, the fighting moves to Villers-Bocage.

While the armour seeks a way forward, the infantry of 1/7th Battalion of The Queen's Royal Regiment prepares to hold Villers-Bocage against the inevitable counterattack. The railway yard is secured by A Company, while B and C companies take up positions in the town guarding the main approaches. D Company is placed in reserve covering the western flank. The battalion's 6 pdr anti-tank guns are concealed throughout the town. All around the British troopers can be heard the clank of heavy steel tracks advancing down the roads towards the town. The battle is about to begin.

THE TIGERS ATTACK

After conferring with Wittmann and Ritgen, *Obersturmführer* Möbius leads his eight remaining Tiger tanks from *1. Kompanie, 101. Schwere SS-Panzerabteilung* towards Villers-Bocage supported by some of his own infantry. Ritgen takes the tanks of *Panzer Lehr* off to the flank. The Germans advance quickly, but in a fragmented manner, with small groups of tanks moving along streets of the town centre and the railway station area.

In the close confines of the streets Tiger, Panzer IV, Cromwell and Firefly tanks all stalk one another in a furious game of cat and mouse. The longer-ranged guns of the German Panzer IV and Tiger tanks are largely negated by the close-in fighting, giving the agile British Cromwell tanks a chance.

The light 6 pdr anti-tank guns engage the heavy Tiger tanks from the sides and rear as they pass. At this very close range, even their small 57mm round, with a bit of luck, can penetrate a Tiger. One by one the large cats are damaged and abandoned, left behind until recovery crews can retrieve them.

British infantry move about in the buildings, waiting for the chance to ambush tanks with their dreaded (or is that dreadful) PIAT anti-tank projectors. The tranquil French town of Villers-Bocage becomes a killing field of great steel behemoths. Even the dreaded Tigers are not immune to death in this place of hewn stone and timber.

GERMAN PANZER LEHR DIVISION

After finishing off the British tanks on Point 213, *Obersturmführer* Rolf Möbius led an ad-hoc group of eight Tigers and supporting troops from the *101. Schwere SS-Panzerabteilung* down towards Villers-Bocage. *Obersturmführer* Wittmann had bought them valuable time with his actions. While the true strength and intentions of the British forces in Villers-Bocage are unknown, they must be dislodged from the town to free the main supply route for the *Panzer Lehr* Division.

The division was well aware of the problem on their flank. All morning the armoured cars had been watching. Even before *Obersturmführer* Wittmann arrived at the headquarters of *Panzer Lehr*, troops were moving to deal with the threat. But with all of their Panzergrenadiers committed against the British 50th Division to their front, the only force available is a few Panzer IV H tanks and a small band of infantry.

Major Helmut Ritgen, deputy commander of *II. Panzerabteilung*, collected every available tank and headed south to seal off the east and west exits out of Villers-Bocage. Running into a screen of heavy anti-tank guns, he quickly realised that the British were there in force. Ritgen ordered his tanks to break off and concentrate for a thrust into the centre of Villers-Bocage to evict the British.

Major Ritgen meets briefly with *Obersturmführer* Möbius to plan their joint attack. The plan is simple. Both companies will send a small force south to the railway station area to clear the British forces there. Meanwhile, the main battle-group will thrust straight for the town centre using speed and shock to compensate for their lack of infantry and throw the British forces out of the town.

The battle for the town of Villers-Bocage has begun. Only time and your skill will determine the victor.

German Panzer Lehr Division

5. Kompanie, 130. Panzerregiment	**Panzer Lehr Kompanie (Confident Veteran) (Page 98 Earth & Steel)**
Kompaniegruppe	Company HQ (two Panzer IV H tanks)
1. Zug, 5. Kompanie	Panzer Platoon (three Panzer IV H tanks)
2. Zug, 5. Kompanie	Panzer Platoon (three Panzer IV H tanks)
3. Zug, 5. Kompanie	Panzer Platoon (three Panzer IV H tanks)
II/130. Aufklärungszug	Motorised Scout Platoon (at full strength, HQ is not equipped with Panzerfaust)
1. Kompanie, 101. s SS-PzAbt	**Schwere SS-Panzerkompanie (Fearless Veteran) (Page 140 Earth & Steel)**
Obersturmführer Möbius	Company HQ (one Tiger I E tank)
1. Zug, 1. Kompanie	Schwere SS-Panzer Platoon (two Tiger I E tanks)
2. Zug, 1. Kompanie	Schwere SS-Panzer Platoon (two Tiger I E tanks)
Aufklärungszug	Armoured SS Scout Platoon (at full strength, HQ is not equipped with Panzerfaust)

BRITISH 7TH ARMOURED DIVISION

With a long list of battle honours, The Queen's Royal Regiment was a source of pride for those that served in its ranks. For the 1/7th Battalion, yet another honour would be added for their part in the fight for Villers-Bocage.

The 1/7th Battalion, The Queen's Royal Regiment, following the advancing 4th County of London Yeomanry, set about securing Villers-Bocage. At the same time tanks of 4 CLY attempted to link up with A Squadron on Point 213. Each attempt was rebuffed by the intervening enemy forces.

In a final act of desperation, Lieutenant Cotton of 4 Troop, B Squadron was sent to find a route around Villers-Bocage. 4 Troop slowly made its way down narrow winding roads anxiously seeking a way through. The rail line to the south-east blocks their way and forces them north, back into Villers-Bocage.

In the meantime, Lieutenant Colonel Desmond Gordon, commanding 1/7 Queen's, positioned his troops. A Company took responsibility for the railway station in the south with C Squadron, 4 CLY. B and C Companies set up with the anti-tank platoon in the solid buildings of the town centre, supported by B Squadron. D Company was placed in reserve holding the western part of the town against German attempts to outflank them.

As the advancing Germans push into the town from the east, Lieutenant Cotton returns to the centre of the town, taking up ambush positions with the infantry. The British are ready and waiting.

For the next six hours the battle rages as the two forces clash in the narrow streets and buildings of the town.

British 7th Armoured Division

B Company, 1/7th Queen's	Lorried Rifle Company (Reluctant Veteran) (page 108 Turning Tide)
B Company HQ	Company HQ (at Full Strength)
10 Platoon, B Company	Lorried Rifle Platoon (at Full Strength)
11 Platoon, B Company	Lorried Rifle Platoon (at Full Strength)
12 Platoon, B Company	Lorried Rifle Platoon (at Full Strength)
5 (Anti-tank) Platoon	Lorried Anti-tank Platoon, (with four 6 pdr guns and Sergeant Tom Stanley)
Battery B , 5th Royal Horse Artillery Regiment	Field Battery (SP), Royal Artillery (with four Sexton self-propelled guns, OP Sherman tank)
B Squadron, 4th CLY	**Armoured Squadron (Reluctant Veteran) (page 84 Turning Tide)**
B Squadron HQ	Company HQ (with two Cromwell IV tanks and two Cromwell VI CS tanks)
1 Troop, B Squadron	Armoured Platoon (with three Cromwell IV tanks and one Firefly VC tank)
2 Troop, B Squadron	Armoured Platoon (as 1 Troop)
3 Troop, B Squadron	Armoured Platoon (as 1 Troop)
4 Troop, B Squadron	Armoured Platoon (as 1 Troop)
Recce Troop	Recce Patrol (with three Stuart V tanks)

THE BATTLE FOR VILLERS-BOCAGE

Villers-Bocage sits upon a major road hub. This road hub contains one of the only north-south roads in the area, while the east-west roads lead directly to Caen in one direction and Avranches in the other. This road structure is critical to supplying the German troops engaged with the British.

This scenario is designed to allow 2 to 4 players to have a game with a larger number of points than normal So get together with some friends and have a blast!

The Villers-Bocage scenario uses the **Ambush** (see page 266 of the rulebook) and **Meeting Engagement** (see page 264 of the rulebook) special rules.

Your Orders

German

You must clear the village of Villers-Bocage in order to remove the threat of allied action against the rear area of Panzer Lehr Division. This will also secure the supply lines being used by units in the sector. Failure is not an option.

British

You must hold the town so that the attack may be renewed at a later date. Inflict as many casualties on the enemy as possible while limiting your losses.

Preparing for Battle

1. Set up the terrain as shown on page 61. Place Objectives at the indicated points. The table is 4'/120cm x 6'/180cm.

2. The British player sets up their force in area (A) within 36"/90cm of the table edge. 4 Troop and 5 (Anti-tank) Platoon may be held in Ambush.

3. The German player now deploys their forces in area (B) within 12"/30cm of the table edge.

Beginning the Battle

1. All reconnaissance platoons make their Reconnaissance Deployment moves starting with the British player.

2. The German force has the first turn.

Ending the Battle

Unlike normal missions, losing one company from your force does not end the game. The other company will continue the battle until it is also Destroyed.

The battle ends:

- At the start of the British turn 11, or
- Whenever one side has both companies Destroyed.

Deciding Who Won

The player or group of players who controls the most objectives wins. If neither player controls any objectives, or they control one objective each, then the game is a draw with neither side able to gain control of the town.

Historical Outcome

Wittmann would become famous all across Germany for his destruction of the British column and the stopping of the Desert Rats. His company on Point 213 would soon be joined by Möbius' 1. Company and together they would destroy A Squadron 4 CLY upon the point. Soon afterwards the Panzer IV tanks and 1 Company would advance into Villers-Bocage. This time the British forces would be waiting. After a hard fight and three more destroyed Tigers the Germans would retreat. Yet as more German forces arrived the pressure upon the defenders grew. Finally, the British pulled back out of Villers-Bocage, not to set foot in the town again for many weeks.

VILLERS-BOCAGE CAMPAIGN

This is the last battle in the Villers-Bocage Campaign. Whichever side wins this battle wins the campaign. The outcome of the Battle for Point 213 scenario has an effect on the forces available for this battle.

If the Germans won the Battle for Point 213, then the deployment area of the British force is halved in this scenario. They must deploy within 18"/45cm of the table edge. The Panzers have raced forward putting pressure on the British forces as they enter the town.

If the British won the Battle for Point 213, then the Germans have lost the race and the entire Schwere SS-Panzerkompanie is placed in Reserve instead of deploying on the table, allowing the British build up to continue before the Germans can muster all of their force. The Schwere SS-Panzerkompanie enters from their deployment zone board edge.

OTHER TIMES AND PLACES

Once again, one could fight this battle with other forces. If the US 2nd Armored Division had made the dash to Villers-Bocage, they would have been fighting with a Tank Company equipped entirely with M4A1 Sherman tanks with 75mm guns and an Armored Rifle Company supported by a weak Armoured Field Artillery Battery.

If you decided to fight this battle on the Eastern Front, the Soviet force could be a Tank Batalon and a Motostrelkovy Batalon supported by a Guards Heavy Tank Company or a Guards Heavy Assault Gun Company.

Of course, you could also try the battle with the roles reversed. Take an allied tank force and attack into a mixed German tank and infantry force. Anything is possible!

Terrain

The battle at Barkmann's Corner was fought in the heart of the Bocage country. The dominant terrain is the Bocage hedgerows. Their six foot high banks are topped with thick bushes and trees creating an almost impassable obstacle to tanks.

Narrow Roads

Both the main road and the side roads are narrow roads. They are too narrow for two tanks to pass, or even to shoot past each other. With a knocked-out tank blocking the road ahead of the column, the only option for the Americans is to force their way through the Bocage hedgerows into the adjacent fields and seek to outflank the Panther.

Bocage Hedgerows

All hedgerows on the table are Bocage hedgerows. These are taller than a tank, Very Difficult Going, and provide Bulletproof Cover to troops sheltering behind them. Teams must start adjacent to a Bocage hedgerow to cross it.

The full Bocage rules are provided in more detail on pages 58 to 59 of the rulebook.

713
711
204

800

STRIKE FOR THE SEA

When informed of the landings, the commander of *84. Armeekorps* (84th Army Corps), General Erich Marcks, wanted *21. Panzerdivision* to attack east of the Orne against the British airborne landings. But as the strategic picture became clearer, stopping the amphibious assault that was quickly gaining a beachhead became the priority.

Only part of the division was thus sent to attack east of the Orne, a *Kampfgruppe* (Battlegroup) under Major von Luck containing most of his *125. Panzergrenadierregiment*, one of the division's five Panzer companies, assault guns, the reconnaissance battalion, and supporting elements. However von Luck was driven back by the defending British paras calling on support from the Royal Navy.

The task of breaking through to the sea fell to the remainder of the division. *Kampfgruppe* Oppeln, comprising his own *22. Panzerregiment*, part of *I/125. Panzergrenadierregiment*, the self-propelled guns of *III/155. Panzerartillerieregiment*, and a company of pioneers, would attack on the right.

Kampfgruppe Rauch with *192. Panzergrenadierregiment* (less the battalion still detached to *716. Infanteriedivision*), the self-propelled guns of *II/155. Panzerartillerieregiment*, and most of the divisional pioneers would attack to Oppeln's left, aiming for Lion-sur-Mer and the radar station at Douvres-la-Délivrande.

As the invaders poured inland, further time was lost in reorganising the men who now had to re-trace their steps through Caen, now choked with rubble following naval and air bombardment, which also created hordes of refugees.

Oppeln's tanks did not begin to deploy near the village of Lebisey until 1600hrs. Frustrated with what he saw as *Generalleutnant* Feuchtinger's delay, General Marcks arrived to take charge of the deployment of the *Kampfgruppen*. He remarked to Oppeln 'If you don't succeed in throwing the British into the sea, we will have lost the war.'

Just how hard this was to be was quickly demonstrated. Oppeln's first objective was the high ground of the Périers Ridge, thought to be German held. It was not. His spearhead ran headlong into troops of 185 Brigade Group, British 3rd Division, who were pressing inland from the beaches.

The 2nd Battalion, The King's Own Shropshire Light Infantry were in the process of attacking Lebisey themselves with the support of the tanks of the Staffordshire Yeomanry. Hearing tanks they prepared for the onslaught, siting their anti-tank guns carefully. They held their fire until the first wave of 40 Panzer IV tanks came into view and then opened a devastating fire, quickly destroying sixteen of the attacking tanks. The Panzers tried to move around the British flank, but running into more British troops, ground to a halt.

Further to the left however, *Kampfgruppe* Rauch found the gap between the Canadian and British beachheads. The battlegroup forced their way through to the coast, linking up with troops from *736. Grenadierregiment* still holding coastal positions to the west of Lion-sur-Mer. This breakthrough caused considerable alarm among the Allied command who rushed troops to deal with the breakthrough. Rauch's men had no sooner reached the coast than they heard and saw masses of gliders flying overhead. Fearing he was about to be cut off, Rauch ordered a withdrawal, giving the initiative back to the Allies.

Overnight, Rauch's men, along with the remnants of Oppeln's group, dug in north of Caen. They had denied the city to the initial thrusts to take it, but their own strike to the sea had amounted to little. With *Panzer Lehr* and *12. SS-Panzerdivision* failing to arrive on D-Day, the planned hammer blow against the landings had fizzled. *21. Panzerdivision* was down to 70 of the 104 tanks with which it had begun the day—losses that it could ill afford with the Allies landing another 300 tanks the next day. The Allies were ashore to stay.

KAMPFGRUPPE OPPELN ATTACKS

As the afternoon of D-Day wears on, the British forces are nearing Caen. One last village stands in their way. As they launch their assault, the German counterattack begins.

The Kampfgruppe Oppeln Attacks scenario uses the **Airborne at Dusk** (below), **Ambush**, **Delayed Reserves**, **Meeting Engagement**, **Reserves**, and **Scattered Reserves** (see pages 264 to 269 of the rulebook) special rules.

YOUR ORDERS

GERMAN

The Allies have landed. High command has finally released you to counterattack and push them back into the sea. You must punch through the leading British troops and breakthrough to the beaches.

BRITISH

Progress toward Caen was going well. Until your lead companies bumped into dozens of counterattacking German tanks. You must hold your gains and push on to Lebisey Wood, ready to push into Caen tomorrow.

PREPARING FOR BATTLE

1. Set up the terrain as shown on page 71. Place the objectives at the indicated points. The table is 4'/120cm by 6'/180cm.

2. The German player places their starting troops at the indicated points. The remainder of their force is held in Reserve.

3. The British player places their starting troops at the indicated points. 5 (Anti-tank) Platoon and 41 Battery are held off the table in Ambush. The remainder of the force is in Delayed and Scattered Reserve.

BEGINNING THE BATTLE

1. The British player has the first turn.

2. This is a mobile battle, so for the British player's first turn of the game all teams (including those deploying from ambush) are considered to be moving.

ENDING THE BATTLE

The battle ends when:

- Either player starts their turn in possession of one of their objectives, or
- Dusk falls under the Airborne at Dusk special rule, and both sides break off the battle.

DECIDING WHO WON

If either player holds one of their objectives, they win the battle. Otherwise the outcome is inconclusive as both sides retire for the few hours of darkness before making another effort in the morning.

AIRBORNE AT DUSK

As dusk fell hundreds of gliders flew over the battlefield, reinforcing the 6th Airborne Division. Fearing being cut off, *General* Feuchtinger ordered the Panzers to break off.

At the start of the British player's turn 7, after taking a Company Morale Check if necessary, the British player rolls a die. On a roll of 5+, dusk falls and the game ends. Otherwise, the German player rolls two dice in their turn, also needing a 5+ on either die to end the game. Each player continues to roll in turn, adding one die each time, until they roll a 5+ and the game ends.

D-DAY MINI CAMPAIGN

Kampfgruppe Oppeln Attacks can be played as part of a mini campaign along with the Kampfgruppe Rauch Attacks scenario on page 72.

The winner of the Kampfgruppe Rauch Attacks scenario wins the campaign. However, your chances of winning that battle will be greatly enhanced or reduced by your performance in this battle.

OTHER PLACES, OTHER TIMES

As it turned out, the only significant armoured force to launch a counterattack on D-Day was *Kampfgruppe Oppeln* of *21. Panzerdivision*. Scattered German tank units did counterattack against the Americans on Utah Beach, but never in more than platoon strength.

You could fight this battle with a US Rifle Company, or even better an Assault Company from *Bloody Omaha*, taking the part of the British. Replace both Armoured Platoons with a single Tank Platoon of three M4 Sherman tanks and the M10 self-propelled guns with towed 3" guns.

To make the game interesting, you could give the Germans all of the tanks available to oppose the landings on Utah Beach. Equip the German counterattack force with obsolete French tanks being used for training. So, instead of Panzer IV tanks, the German Panzer Platoons have Panzer 39H(f) Hotchkiss tanks with a Panzer III tank as the command vehicle.

The Germans debated moving their *12. SS-Panzerdivision* into a position opposite Omaha Beach. You could also play the scenario with the same American force facing off against a counterattack by fearless SS-Panzer Platoons equipped with Panzer IV tanks.

Moving further afield, the Red Army conducted many river crossing operations. These often faced German counterattacks trying to push them back to the river line. You could play the scenario with a Soviet Strelkovy Batalon in place of the British force.

German Kampfgruppe Oppeln

22. Panzerregiment	Panzerkompanie (Confident Veteran) (Page 52 Earth & Steel)
Oberst von Oppeln-Bronikowski	Company HQ (with two Panzer IV H tanks at A)
5. Panzerkompanie	Panzer Platoon (with five Panzer IV tanks at A)
1./125. Panzergrenadierregiment	Panzergrenadier Platoon (at full strength, with no Panzerfaust anti-tank launchers and no trucks, at B)
Verstärkungen	**Reserves**
2. Panzerkompanie	Panzer Platoon (with five Panzer IV tanks)
3. Panzerkompanie	Panzer Platoon (with five Panzer IV tanks)
III/155. Panzerartillerieregiment	Armoured Artillery Battery (with six 10.5cm (Sf) Lorraine Schlepper self-propelled guns)

Oberst Hermann von Oppeln-Bronikowski was a cavalryman since before the First World War, although he won his Iron Cross fighting as an infantryman in that war. His service in Poland, France and Russia won him a second Iron Cross and a Knight's Cross. After being wounded at Kursk, he took command of 22. Panzerregiment. On D-Day he was given the task of striking for the beaches with the only Panzer force available to intervene. With his *4. Kompanie* fighting parachutists across the Orne and *1. Kompanie* delayed, it was the composite *5. Kompanie* from *II Panzerabteilung* that led the way. The remainder of the regiment followed as quickly as it could, pushing towards Lion-sur-Mer on the beach.

British 185 Brigade Group

2nd Battalion, Kings Own Shropshire Light Infantry	Rifle Company (Confident Trained) (Page 44 Turning Tide)
Lieutenant Colonel Maurice	Company HQ (with Universal Carrier)
W Company	Rifle Platoon (at full strength at W)
X Company	Rifle Platoon (at full strength at X)
Y Company	Rifle Platoon (at full strength at Y)
C Squadron, Staffordshire Yeomanry	Independent Armoured Platoon (rated Confident Veteran with three Sherman III tanks and one Firefly VC tank at C)
Flank Guard	**Ambush**
5 (Anti-tank) Platoon	Anti-tank Platoon (with four 6 pdr guns)
41 Anti-tank Battery (SP), 20th Anti-tank Regiment	Assault Anti-tank Battery (SP), Royal Artillery (with two M10 (3") SP)
Reserves	**Delayed and Scattered Reserves**
Z Company	Rifle Platoon (at full strength)
A Squadron, Staffordshire Yeomanry	Independent Armoured Platoon (as above)
7th Field Regiment, Royal Artillery	Assault Field Battery (SP), Royal Artillery (with eight Priest self-propelled guns)

Lieutenant Colonel Maurice's Kings Own Shropshire Light Infantry led the push by 3rd Division's 185 Brigade to Caen on the afternoon of D-Day. After taking Beuville and Bieville, W and Y Companies (the KSLI used W, X, Y, and Z rather than the more usual A, B, C, and D for their companies) set out for Lebisey. This village was reported to be lightly held, however the Panzergrenadiers of *125. Panzergrenadierregiment* had just arrived there prior to their own attack. After halting at a tank obstacle, the Shropshires were preparing for another attack just as the first German tanks appeared over the ridge. Their tanks and anti-tank guns destroyed the leading German tanks as more approached.

Terrain

The terrain behind Sword Beach is open fields. There are no fences or hedges to break up the wide open spaces. Even the crops are still young, barely knee high, giving no cover.

The low ridges are tall enough to hide a tank, with gentle slopes that provide no obstacle to movement.

However, the wooded stream between Bieville and Lebisey has been transformed into a tank obstacle for German training exercises. It is rated as Very Difficult Going.

The villages of Beuville, Bieville and Lebisey are set amongst orchards and woods and bounded by Bocage hedgerows. The hedgerows are tall, far too tall to be seen over from the ridges, and gave the villages the appearance of woods from a distance.

The roads were wider than in the western part of Normandy and allow vehicles to manoeuvre normally and pass other vehicles if need be.

Bocage Hedgerows

All hedgerows on the table are Bocage hedgerows. These are taller than a tank, Very Difficult Going, and provide Bulletproof Cover to troops sheltering behind them. Teams must start adjacent to a Bocage hedgerow to cross it. The full Bocage rules are provided in more detail on pages 58 to 59 of the rulebook.

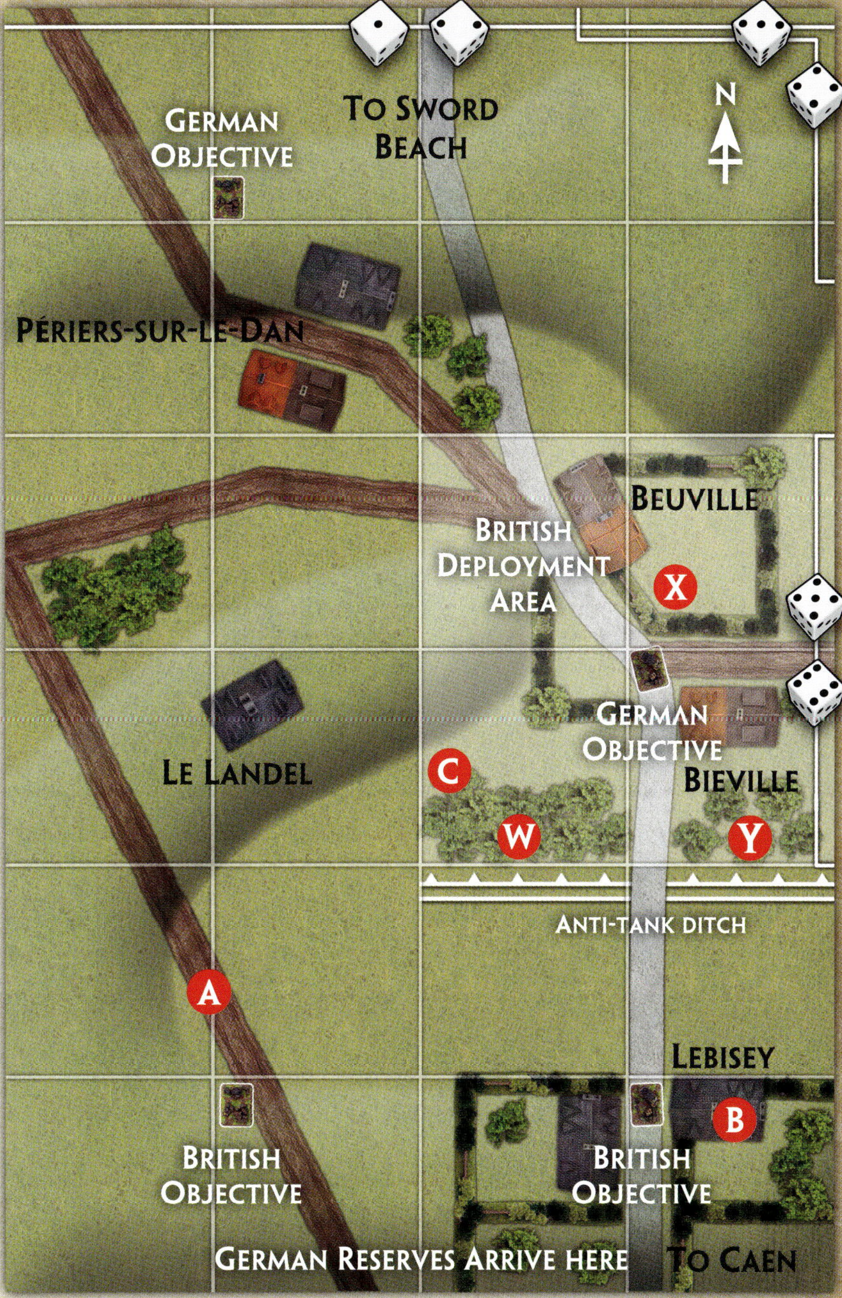

10

KAMPFGRUPPE RAUCH ATTACKS

Soon after the tanks of *Kampfgruppe Oppeln* attacked, the *Panzergrenadiere* of *Kampfgruppe Rauch* launched their own attack on their western flank. Finding the gap between the British and Canadian beaches, they reached the radar station at Douvres-la-Délivrande before encountering opposition.

The Kampfgruppe Rauch Attacks scenario uses the **Airborne at Dusk** (see page 69), **Reserves**, and **Meeting Engagement** (see pages 264 and 268 of the rulebook) special rules.

YOUR ORDERS

GERMAN

You must reach the sea at Lion-sur-Mer, splitting the Allied invasion force in two. The division will then turn and thrust along the beaches pushing the invaders back into the sea.

CANADIAN

You have made it ashore with surprisingly light casualties. Progress inland has been delayed by massive traffic jams on the beaches and sporadic German opposition. You must strike inland towards Caen, clearing the strongpoint at Douvres-la-Délivrande on the way.

PREPARING FOR BATTLE

1. Set up the terrain as shown on page 73. Place the objectives at the indicated points. The table is 4'/120cm by 6'/180cm.

2. The Canadian player places their starting troops in their deployment area. The remainder of their force remains in Reserve. These will arrive between Tailleville and Beny-sur-Mer.

3. The German player places their troops in the indicated areas.

BEGINNING THE BATTLE

1. The Canadian player has the first turn.

2. This is a mobile battle, so for the Canadian player's first turn of the game all teams are considered to be moving.

ENDING THE BATTLE

The battle ends when dusk falls under the Airborne at Dusk special rule, and both sides break off the battle.

DECIDING WHO WON

The German player gets one victory point for holding the Douvres-la-Délivrande radar station objective and two victory points for holding the Lion-sur-Mer objective.

The Canadian player gets one victory point for holding the Douvres-la-Délivrande radar station objective and one victory point each for holding the Colomby-sur-Thaon and Cazelle objectives.

The player holding the most victory points at the end of the game wins.

D-DAY MINI CAMPAIGN

Kampfgruppe Rauch Attacks can be played as part of a mini campaign along with the Kampfgruppe Oppeln Attacks scenario on page 69. Play the Kampfgruppe Oppeln Attacks scenario first, then this scenario. The winner of the Kampfgruppe Rauch Attacks scenario wins the campaign.

If the German player won the Kampfgruppe Oppeln Attacks scenario, the victorious tanks of *Kampfgruppe* Oppeln arrive to assist *Kampfgruppe* Rauch. The German player receives a Panzer Platoon with three Panzer IV tanks as a Delayed Reserve arriving at the edge of the German deployment area C behind Douvres-la-Délivrande in this scenario.

If the British won the Kampfgruppe Oppeln Attacks scenario, the attack by *Kampfgruppe* Rauch is delayed and the German forces that would normally start in Douvres-la-Délivrande in area C start at Cazelle in area E instead.

OTHER PLACES, OTHER TIMES

Like the Kampfgruppe Oppeln Attacks scenario, you could easily fight this battle with other forces to represent different battles or possibilities. A perfect example would be the planned counterattack against Omaha Beach by *Kampfgruppe Meyer* (See *Earth & Steel* page 32 for details on *352. Infanteriedivision*). As it turned out they spent all day marching in the division's rear, but things could have been very different if they had been ordered to counterattack the beach.

Use a US Rifle Company, or even better an Assault Company from *Turning Tide*, in place of the Canadian force. Replace both Armoured Platoons with a single Tank Platoon of three M4 Sherman tanks and the M10 self-propelled guns with towed 3" guns.

Use *Earth & Steel* to create *Kampfgruppe Meyer* (see page 32 of *Earth & Steel*) with a StuG Abteilung with supporting Fusilier Platoons and a Festungs Artillery Battery.

The sort of confused breakthrough situation portrayed in Kampfgruppe Rauch Attacks happened many times on the Eastern Front. You could replace the Canadian force with a Soviet Motostrelkovy Batalon that has forced a river and is breaking through into the depths of the German defences with only a small 'Hedgehog' position between them and victory. The small German Panzergrenadier force is attempting to reach the river crossings to stem the tide.

Terrain

This battle takes place along a broad north-south ridge. A stream drains east cutting down the side of the ridge. In early summer the stream is shallow, barely an obstacle, and counts as Difficult Going.

In the early summer the crops are low and provide little in the way of concealment, giving the defenders a wide-open field of fire, although they do provide good going for the attacking forces.

Only the low bed of the railway line provides much of an obstacle, but even that is just Difficult Going and provides no concealment, although it is Bulletproof Cover for stationary Infantry and Man-packed Gun teams immediately behind it.

Douvres-la-Délivrande

The night-fighter control radar station at Douvres-la-Délivrande is well fortified. The first line of defence is a minefield. Inside that is a barbed wire entanglement. As normal, each section of defences is usually 8"/20cm long by 2"/5cm wide. The exception to this is the barbed wire entanglements where each side of the fortification is made up of a 8"/20cm and a 4"/10cm section.

The garrison is well provided with Trench Lines for protection. They are protected by an HMG Nest, a 5cm PaK38 Nest, and a 2cm FlaK38 Nest (see pages 217 to 222 of the rulebook for the rules on Nests).

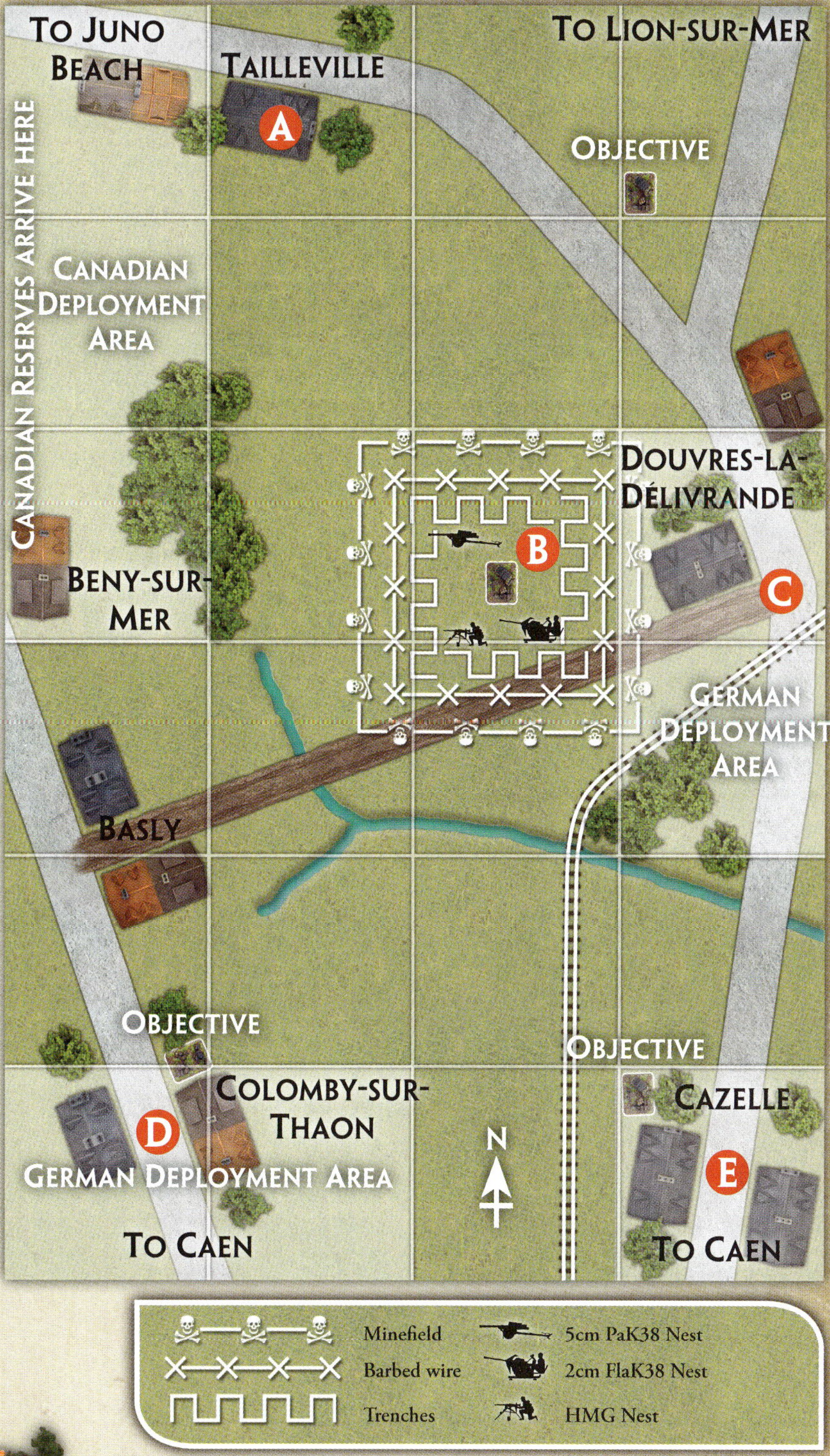

GERMAN KAMPFGRUPPE RAUCH

Originally serving in the army between the wars, *Oberst* Rauch rejoined the army in 1935. After commanding *Pionier* and *Panzergrenadier* units in combat, he took over command of the newly formed *192. Panzergrenadierregiment*. On D-Day his command was deployed as a reserve for *716. Infanteriedivision* manning the coastal defences.

Counterattacking with his armoured I Battalion, *Kampfgruppe* Rauch reached the night-fighter control station at Douvres-la-Délivrande, defended by a small force from *736. Grenadierregiment* and the *Luftwaffe* (Air Force) signals troops manning it, before pushing on to the sea at Luc-sur-Mer.

German Kampfgruppe Rauch

192. Panzergrenadierregiment	Gepanzerte Panzergrenadierkompanie (Confident Veteran) (Page 56 Earth & Steel)
Oberst Rauch	Company HQ (with artillery Observer team, at C)
I/192. Panzergrenadierregiment	Gepanzerte Panzergrenadier Platoon (at full strength, at C)
II/192. Panzergrenadierregiment	Panzergrenadier Platoon (with two squads and no trucks, at D)
4/192. Panzergrenadierregiment	Armoured Anti-tank Platoon (with three S304(f) (PaK40) half-tracks, at C)
II/155. Panzerartillerieregiment	Armoured Artillery Battery (with three 15cm (Sf) Lorraine Schlepper self-propelled guns, at E)
3/53. Luftwaffe Nachrichtenregiment	Luftwaffe Field Platoon (rated Reluctant Trained with two squads, at B) with Gun teams in Open Bunkers as shown on the map
Stab/II/736 Grenadierregiment, 716. Infanteriedivision	Two independent SMG teams (rated Reluctant Trained at A)

No command teams have Panzerknacker training or Panzerfaust anti-tank launchers.

8 CANADIAN BRIGADE GROUP

8 Canadian Brigade Group under Brigadier Blackadder landed on the eastern half of Juno Beach. By midday the North Shore Regiments reported Tailleville clear and were set to move onto their final objective—the radar station. They did not set out until 1730hrs, though, as fighting flared up repeatedly against elements of the headquarters of *736. Grenadierregiment*. The Queen's Own Rifles supported by the Fort Garry Horse pushed inland to Beny-sur-Mer and Basley. About this time the follow-on 9 Brigade Group finally escaped the massive traffic snarl-up on the beach and moved inland towards Carpiquet airfield at Caen, with the North Novas and the tanks of the Sherbrookes in the lead.

8 Canadian Brigade Group

8 (Eastern) Brigade Group	Canadian Rifle Company (Confident Trained) (Page 44 Turning Tide)
Brigadier Ken Blackadder	Company HQ (with Universal Carrier)
The Queen's Own Rifles of Canada	Rifle Platoon (with two squads)
The North Shore (New Brunswick) Regiment	Rifle Platoon (with two squads)
The Cameron Highlanders of Ottawa (MG)	Machine-gun Platoon (one section in MMG Carriers)
10th Armoured Regiment (The Fort Garry Horse)	Independent Armoured Platoon (three Sherman III tanks)
3rd Anti-tank Regiment, Royal Canadian Artillery	Assault Anti-tank Battery (SP), Royal Artillery (with two M10 (3") SP)
14th Field Regiment, Royal Canadian Artillery	Assault Field Battery (SP), Royal Artillery (with four Priest self-propelled guns)
9 (Highland) Brigade Group	**Reserves**
The North Nova Scotia Highlanders	Rifle Platoon (at full strength)
27th Armoured Regiment (The Sherbrooke Fusiliers)	Independent Armoured Platoon (with three Sherman III tanks and one Firefly VC tank)

11
OUT OF LUCK

At 0745hrs on 18 July, 1944, 3rd Division began its part of Operation Goodwood. Initial progress was fast, the German defenders stunned after hours of heavy bombing and shelling, with Sannerville being captured around midday. The next objective was the chateau at Banneville-la-Campagne.

The Out of Luck scenario uses the **Disorganised Counterattack** (see page 78), **No Retreat** (below), and **Prepared Positions** (see page 264 of the rulebook) special rules.

YOUR ORDERS

BRITISH

Capture and clear Banneville-la-Campagne and the surrounding area. Hold this position as the flanking brigades exploit to Troarn and Manneville.

GERMAN

Halt the British attack at all costs. Do not give up your positions. You must protect the artillery positions at Banneville-la-Campagne and prevent a British breakthrough.

PREPARING FOR BATTLE

1. Set up the terrain as shown on page 77. The table is 4'/120cm by 6'/180cm.
2. The German player places their starting troops in their deployment area. The Heavy Anti-tank Platoon and Sniper teams are held off the table in Ambush. The rest of their force remains off-table in Random Reserve.
3. The German artillery observer teams can deploy anywhere in the German deployment area.
4. The British player deploys their entire force in their deployment area.

BEGINNING THE BATTLE

1. All German troops on the table start the game in Prepared Positions.
2. The British player has the first turn.

ENDING THE BATTLE

The battle ends at nightfall at the start of British turn 10 after making their Company Morale Check if necessary.

DECIDING WHO WON

There are three objectives, all buildings. These are: Banneville-la-Campagne, the Chapel, and the House by the Caen-Troarn Road.

The player holding the most objectives at the end of the game wins.

NO RETREAT

The battle around Sannerville-la-Campagne Chateau was very disjointed from the German perspective. The heavy British bombing disrupted communications and left the German troops to fight their own individual battles. Aware of the importance of the battle and Hitler's 'No Retreat' orders, the soldiers fought, and died, where they stood.

The German forces in the Out of Luck scenario always pass Company Morale Checks that they need to take, even though they have no Company Command team.

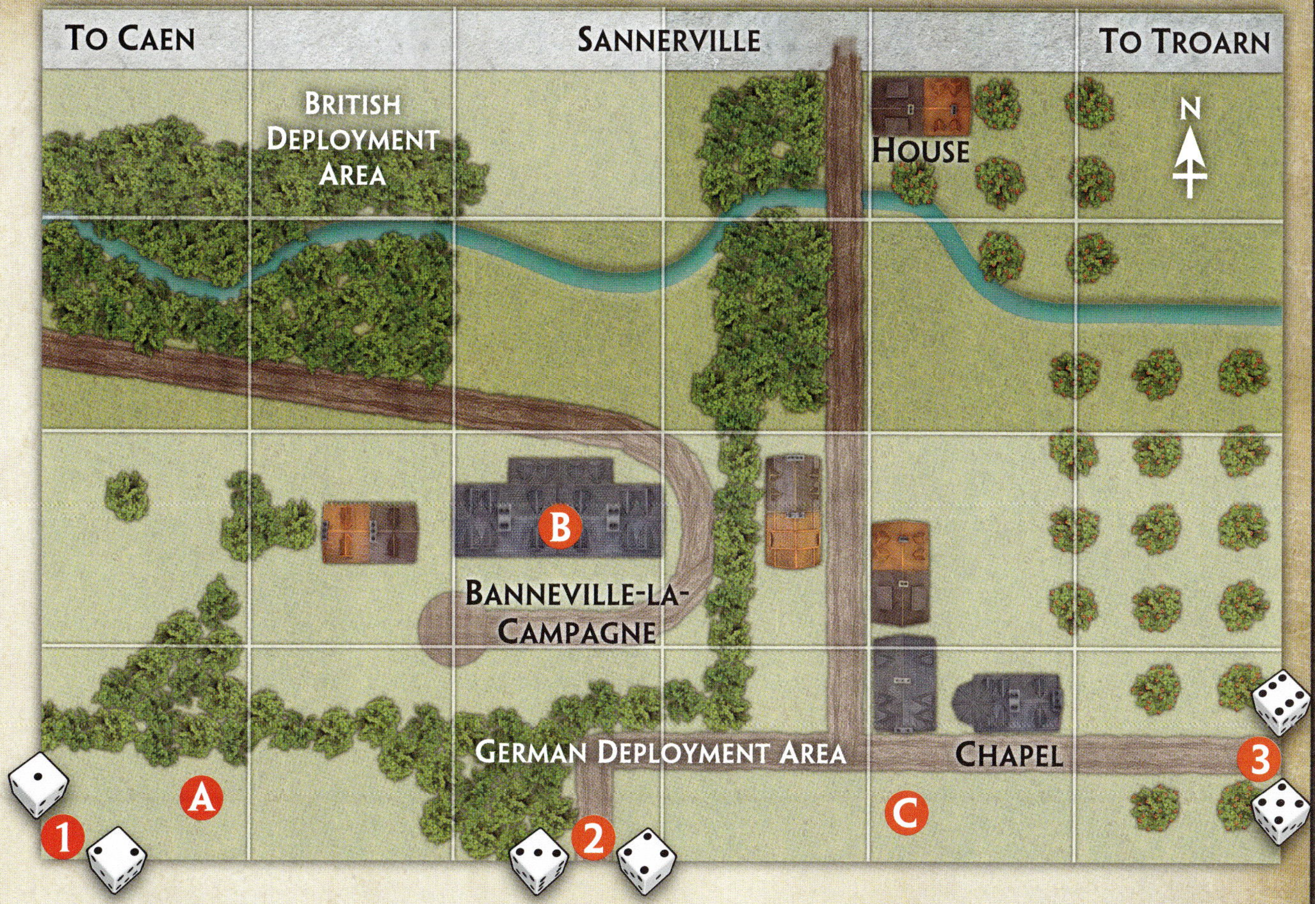

Terrain

As the Suffolks continued their advance southward after capturing Sannerville, they crossed the main east-west Troarn-Caen road approaching the chateau of Banneville-la-Campagne, their final objective for the day.

The Banneville-la-Campagne Chateau is a large building with a two-storey centre and two more rooms making up the east and west wings. Use the Multi-room Buildings rules on pages 24 and 28 of the rulebook for this building.

The stream running across the battlefield is Very Difficult Going and provides Bulletproof Cover and Concealment to stationary infantry in the stream, sheltering against the steep banks.

Although the woods in the western half of the battlefield are badly damaged by bombing and shelling, they still count as normal Woods and are Difficult Going.

The Orchards to the east are more open. While they are area terrain like Woods and count as Difficult Going, they only Conceal teams within them. They are open enough that teams within or beyond the orchard can see and be seen at any distance.

The remaining terrain is open fields of dry grass, covered in thin grey dust from the bombed-out ruins of Sannerville.

German Kampfgruppe Luck

125. Panzergrenadierregiment	Kampfgruppe Luck (Confident Veteran) (see pages 51 to 83 Earth & Steel)
6. Jägerkompanie, 46. Jägerregiment (LW)	Luftwaffe Jäger Platoon (rated Reluctant Trained, with three squads, at B)
2. Batterie, 155. Panzerartillerieregiment	Motorised Artillery Battery (with four 12.2cm FH316(r) howitzers at C)
10. Batterie, 155. Panzerartillerieregiment	R-Vielfachwerfer Battery (with loading crews at A)
Hinterhalt	**Ambush**
Scharfschützen	Two Sniper teams
3 Batterie, 1039. Artillerie Pak Abteilung	Heavy Anti-tank Gun Platoon (with two 8.8cm PaK43)
Verstärkungen	**Disorganised Counterattack**
Major von Luck	Major von Luck (as SMG team with Kübelwagen jeep or in Panzer IV H tank)
7. Kompanie, 125. Panzergrenadierregiment	Panzergrenadier Platoon (with three squads, HQ equipped with Panzerfaust)
2. Kompanie, 125. Panzergrenadierregiment	Gepanzerte Panzergrenadier Platoon (with two squads, HQ equipped with Panzerfaust)
8. Batterie, 52. Flakregiment	Luftwaffe Heavy Anti-aircraft Platoon (rated as Reluctant Trained, with two 8.8cm FlaK36 guns with 8 crew each)
4. Batterie, 200 StuG Abteilung	Beute StuG Platoon (with one 7.5cm (Sf) 39H and two 10.5cm (Sf) 39H)
4. Panzerkompanie, 22. Panzerregiment	Panzer Platoon (with three to five Panzer IV H tanks, roll 1 to 3 = 3 tanks, 4 or 5 = 4 tanks, 6 = 5 tanks)
1. Panzerkompanie, 503. Schwere Panzerabteilung	Schwere Panzer Platoon (with one to three Königstiger tanks, roll 1 to 3 = 1 tank, 4 or 5 = 2 tanks, 6 = 3 tanks)

Kampfgruppe Luck was one of the first units in combat on D-Day. It has been in constant combat since then. Despite being reinforced with part of *16. Felddivision (LW)*, the constant hammering of British artillery and battles fought to contain the bridgehead across the Orne have taken their toll. Most companies are now just platoons.

Disorganised Counterattack

Operation Goodwood penetrated deep into the German defensive lines. Banneville-la-Campagne was an artillery position with reserve tank units hidden in woods just a few thousand metres away, where they suffered heavily from the bombing. The main defences were manned by *16. Felddivision (LW)* backed up by *Oberst* von Luck's *125. Panzergrenadierregiment*. These were supported by 8.8cm anti-tank guns from *1039. Artillerie-PaK-Abteilung* and 8.8cm anti-aircraft guns of *III Flaksturmkorps*.

The German forces counterattacking later in the day came from a wide variety of units. The new Königstiger tanks of *503. Schwere Panzerabteilung* launched several counter-attacks to protect Troarn. The surviving Panzer IV tanks of *21. Panzerdivision* mainly headed south to counterattack the armoured penetration, joining Becker's StuG batteries fighting a running battle on the flanks of the attack.

Starting with their third turn, the German player rolls one die each turn for their reserves. The score on the die determines which platoon arrives from reserve. As an example, a roll of 4 would have a StuG Platoon arrive from reserve.

If the roll is the same as a number that has already been rolled, then take the next lower number if that platoon has not yet arrived, and bring it on now instead. If both the rolled platoon and next lower platoon have already arrived, then no reserves arrive this turn.

Roll another die to determine which point the reserve platoon will arrive. On a roll of 1 or 2 the platoon arrives at 1 in the corner of the table. On a roll of 3 or 4 the platoon arrives at 2 along the southern road. On a roll of 5 or 6 the platoon arrives at 3 along the eastern road.

Major Hans von Luck arrives with the first Reserve platoon to arrive. The German player can choose whether he arrives in the tank or in his Kübelwagen. If he arrives in a tank, then the Panzer Platoon has a maximum of four tanks when it arrives. Unlike normal Reserves, Major Von Luck does not get to re-roll the reserves die roll in this scenario.

8 Brigade Group

1st Battalion, The Suffolk Regiment	Rifle Company (Confident Trained) (Page 44 Turning Tide)
Lieutenant Colonel Gough	Company HQ (with Universal Carrier)
7 Platoon, A Company	Rifle Platoon (with two squads)
8 Platoon, A Company	Rifle Platoon (with two squads)
16 Platoon, D Company	Rifle Platoon (with two squads)
17 Platoon, D Company	Rifle Platoon (with two squads)
7 Platoon, 2nd Bn, The Middlesex Regiment (MG)	Machine-gun Platoon (with two sections in MMG Carriers and PIAT team)
45 Battery, 20th Anti-tank Regiment, Royal Artillery	Assault Anti-tank Battery (SP), RA (with four M10 3" SP)
303 (Fife) Battery, 76th (Highland) Field Regiment, Royal Artillery	Assault Field Battery (SP), Royal Artillery (with eight Priest self-propelled guns)
B Squadron, 13/18th Hussars	**Armoured Squadron (Confident Trained)**
B Squadron, 13/18th Hussars	Company Command and 2iC Sherman III tanks
6 Troop, 13/18th Hussars	Armoured Platoon (with three Sherman III tanks and one Firefly VC tank)
7 Troop, 13/18th Hussars	Armoured Platoon (as above)

The 1 Suffolks had the dubious honour of leading the advance for 8 Brigade. After clearing Sannerville, B and C Companies were left holding the town while the rest of the battalion attacked the defenders of Banneville-la-Campagne with B Squadron of the 13/18th Hussars in support. By this stage most companies were severely weakened by casualties and barely strong enough to field two platoons.

Casualties amongst supporting troops were much lighter, so they could count on plentiful support from the machine-gunners and the Royal Artillery anti-tank and field artillery.

Fortunately the Germans were short of troops too and were unable to mount a strong defence. The Banneville-la-Campagne chateau was cleared in an hour and a half.

British Kangaroos

Lieutenant General Simonds, commanding II (Canadian) Corps, wasn't the first to think of using armoured artillery carriages as armoured personnel carriers (see page 59 of *Turning Tide*), just the first to actually do it.

Lieutenant General Crocker of I Corps had suggested converting Priest self-propelled guns into armoured personnel carriers for Operation Goodwood, but was turned down as the benefits were not perceived to be sufficient. After the problems caused in that operation by the inability of the infantry to keep up with the armoured spearheads, things were rethought for Operation Totalize where the armoured personnel carriers were used.

There is an interesting question as to what might have happened had Crocker got his way. You could add four Armoured Personnel Carrier Sections (see Breaching Group pages 55 to 59 of *Turning Tide*) to the British force for the battle and see what difference it makes.

Other Places, Other Times

While a major breakout eluded the Allies for the first two months of the campaign, small-scale breakthroughs like the 3rd Division's were reasonably common in every operation from Epsom to Bluecoat.

You could play the Out Of Luck scenario with a US Rifle Company attacking to represent one of the small US breakthroughs on the road to St. Lô by the 29th Infantry Division *(see page 218 Turning Tide)*. The battle represents the Americans pushing through the German lines on a narrow sector trying to take a chateau before the German counterattack arrives to stop them.

For this scenario, replace the Machine-gun Platoon with an Ammunition and Pioneer Platoon for the Americans and change the German Königstiger tanks into Panther tanks.

You might also want to change the map to add fields surrounded by Bocage hedgerows around the Orchard and across much of the table apart from the chateau grounds. While this will slow the US advance, it will also limit the German fields of fire and slow down their reserves as well.

12

LEPKOWSKI'S PRISONBREAK

The Brasprats Raid, 16 August 1944

When *Leutnant* (lieutenant) Erich Lepkowski called his *5. Kompanie* (5th Company) together he knew the daring mission he was about to give his men had to succeed. Failure meant certain death for 130 of his fellow *Fallschimjäger* (paratrooper) comrades in arms.

Lepkowski was determined to rescue them, for they would certainly do the same for him. He took command of the rescue mission and without a moment to lose, called for volunteers. He watched with pride as all of his men stepped forward to join him.

Lepkowski outlined the cold, hard details of the situation. Partisans of the French Forces of the Interior (FFI) caught over 130 of their fellow *Fallschirmjäger* troops. A few troopers were able to escape and made the harrowing journey through the FFI and American perimeter and back to German lines.

The soldiers reported that the French were mistreating the prisoners. Unfortunately, reports of this kind were tragically common as the newly empowered French fighters exacted terrible revenge upon German prisoners for the occupation. A captured German's life expectancy in the hands of the French was only a matter of how long they fancied keeping him alive. Sometimes this was only minutes or days, if the German prisoner was lucky.

Lepkowski dismissed his men to prepare for the mission. Meanwhile he planned out how he was going to get his force all the way to where the prisoners were being held, some 30 miles (50km) away in a small village called Brasprats. Somehow, Lepkowski needed to break through the thin American perimeter and through FFI territory, rescue the prisoners before the FFI could execute them, and then get back to Brest as quickly as possible.

The force only had enough men, weapons, and ammunition for a single one-way trip if they were forced to fight on the way in and the way out. The key to success would be to sneak their way into Brasprats, spring the prisoners, and then fight their way back to German lines.

Lepkowski ran around to all of the regimental headquarters in the division and gathered as many of the captured Allied trucks as he could. He had all of the German insignia stripped away, Allied stars painted on, and French liberation slogans scribbled on the sides. The paratroopers then put on French partisan clothing over their uniforms and mounted up in the 'Allied' vehicles.

The genius of Lepkowski was thanks in part to his divisional commander General Hermann-Bernhard Ramcke, who had won a reputation for launching daring and successful raids against the Allies in North Africa. Ramcke was also concerned for his captured men and he authorised Lepowski to use three captured American tanks to bolster the rescue mission's firepower.

Just before the vehicles departed, the paratroopers added more pieces to their disguise. American and French flags were flown from the vehicles to help trick the FFI and American soldiers, but also served to protect them from any enemy aircraft that might fly overhead.

Furthermore, false orders to release the prisoners into the custody of the Lepkowski's 'Americans' were drawn up to help persuade the FFI to hand over the prisoners. Finally, as an added security measure, French-speaking soldiers were selected to drive the vehicles in case they were stopped and questioned.

In the early hours of 16 August Lepkowski gave the order to advance. The column of 18 trucks and three tanks lurched forward to its first challenge, the American perimeter.

The *Fallschirmjäger* launched several diversionary attacks and the Americans dutifully reacted. While the Americans were distracted elsewhere, Lepkowski slipped through the lines and into FFI territory. The column passed through several checkpoints. The FFI did not suspect anything, especially that an entire company of *Fallschimjäger* were hidden in plain sight!

As the column neared Brasprats, the *Fallschimjäger* in the back of the trucks pulled down the canvas coverings over the truck beds, removed their disguises, and readied their weapons. The men sat quietly as they listened to the FFI guards chat with the drivers. Soon the truck was moving again and the column entered the town.

The prisoners were being held in a small school building. Lepkowski unfolded a map with notes and directions scribbled in red pencil. He ordered the column to the school building. As they pulled up to the objective, something went wrong. A rifle shot cracked through the air and with that Lepkowski's ruse was up!

Instantly, the *Fallschirmjäger* troops leapt from the trucks and sprang into action securing the building and the area around it. The prisoners inside had heard their comrades outside and quickly overpowered the guards, which they tied up and took with them.

The two German forces were glad to see each other, but didn't waste a moment beating a hasty retreat out of the village. The highly professional paratroopers easily overcame the untrained FFI. The tanks pelted the FFI guard posts high-explosive shells, knocking out the few heavy weapons the French had.

After about 20 minutes of fighting the *Fallschirmjäger* reached the village roadblock. The French put up a fight and after a sharp battle the Germans swept aside the French partisans and broke through onto the open road.

The column raced toward Brest smashing through the first few FFI roadblocks. However, resistance grew as they got closer to the front line. The disorganised FFI finally managed to reinforce the last FFI roadblock ahead of the *Fallschirmjäger* and another short battle broke out. Once again the tanks led the way and blasted their way through the roadblock. The Germans troops mounted up again and pushed through.

As the column reached the American perimeter they were pleased to discover that the Americans were still occupied by Ramcke's diversionary attacks. The column slipped back into Fortress Brest without any more trouble.

Lepkowski's raid was a terrific success. The column had travelled 75 miles (120 km) round trip. They had rescued all 130 prisoners and even captured 15 FFI soldiers. Lepkowski was impressed by the French fighters and successfully argued to have them interred as regular Prisoners of War, sparing them from the firing squad as guerrilla fighters. Even more amazing was that the German casualties were only three lightly wounded soldiers despite all of the heavy fighting.

Erich Lepkowski was an instant hero. Ramcke promoted him to *Oberleutnant* (1st Lieutenant) and he was awarded the Knight's Cross. He continued to provide excellent service to his beloved *Fallschirmjäger* until he was horribly wounded in the closing days of the battle for Brest. Lepkowski, a breath or two away from death, was rescued from a pile of corpses by a German doctor and treated. In time Lepkowski made a full recovery and went on to serve in Germany's post-war paratrooper force.

SCENARIO FORCES

The Big Lepkowski scenario uses forces from the Normandy books, *Earth & Steel* and *Turning Tide*. You can find the *Fallschirmjäger* force used in this scenario on page 194 in *Earth & Steel*. You can find an FFI Company in *Turning Tide on page* 229.

LEPKOWSKI'S RAIDERS

FALLSCHIRMJÄGER KOMPANIE (FEARLESS VETERAN)

Company HQ:
Company Command and
2iC Command SMG teams

Fallschirmjäger Platoon:
1x Command SMG team, and
9x Rifle/MG teams

Captured Tank Platoon:
1x captured Command Stuart tank, and
2x captured M8 Greyhound Armoured Cars

Captured Truck Platoon: 4x captured 2½-ton trucks

PRISONERS OF WAR

Fallschirmjäger Platoon:
1x Command SMG team, and
9x Rifle/MG teams

THE FREE FRENCH FORCES

FFI BATTALION (CONFIDENT CONSCRIPT)

Battalion HQ:
Company Command and
2iC Command SMG teams

FFI Company:
1x Command SMG team,
15x Rifle teams,
1x Captured HMG team, and
1x Captured Anti-tank gun

RESERVES

FFI Company:
1x Command SMG team, 4x Rifle teams, and
4x SMG teams

FFI Company:
1x Command SMG team, 4x Rifle teams, and
4x SMG teams

SCENARIO SPECIAL RULES

RESERVES

You can find the Reserves special rule on page 268 of the rulebook.

RESERVE CHECKPOINT

There are three Reserve Checkpoints along the road marked X. *French platoons arrive from reserve randomly at one of of these Checkpoints. Roll a dice and look at the scenario map to determine at which Reserve Checkpoint your company will arrive.*

Arriving platoons must be deployed within 12"/30cm of the Checkpoint, and more than 10"/25cm away from all enemy teams.

RUSE

The Fallshcirmjäger have completely fooled the FFI with their ruse and caught them completely unprepared to deal with the elite troop's attack. As such, FFI platoons cannot attempt to Dig In during the course of this scenario.

CAPTURED VEHICLES

Lepkowski commandeered and used several captured Allied tanks to add some firepower to his mission. These were likely M5 Stuarts and M8 Greyhound armoured cars. They were used as assault guns to blast any enemy guns and strongpoints the convoy might encounter.

The M8 Greyhounds in the Captured Tank Platoon are not Reconnaissance teams.

The Captured Truck Platoon follows the rules for Transport Platoons found on pages 17, 47, and 48 of the rulebook.

THE BIG LEPKOWSKI

German paratroopers are being held as Prisoners of War in the village of Brasprats by Free French forces. Past experience has shown that the French are not particularly kind toward their former occupiers, and its only a matter of time before they take their revenge!

The Big Lepkowski scenario uses the **Reserves**, **Reserve Checkpoint**, and **Ruse** (see page 82) special rules.

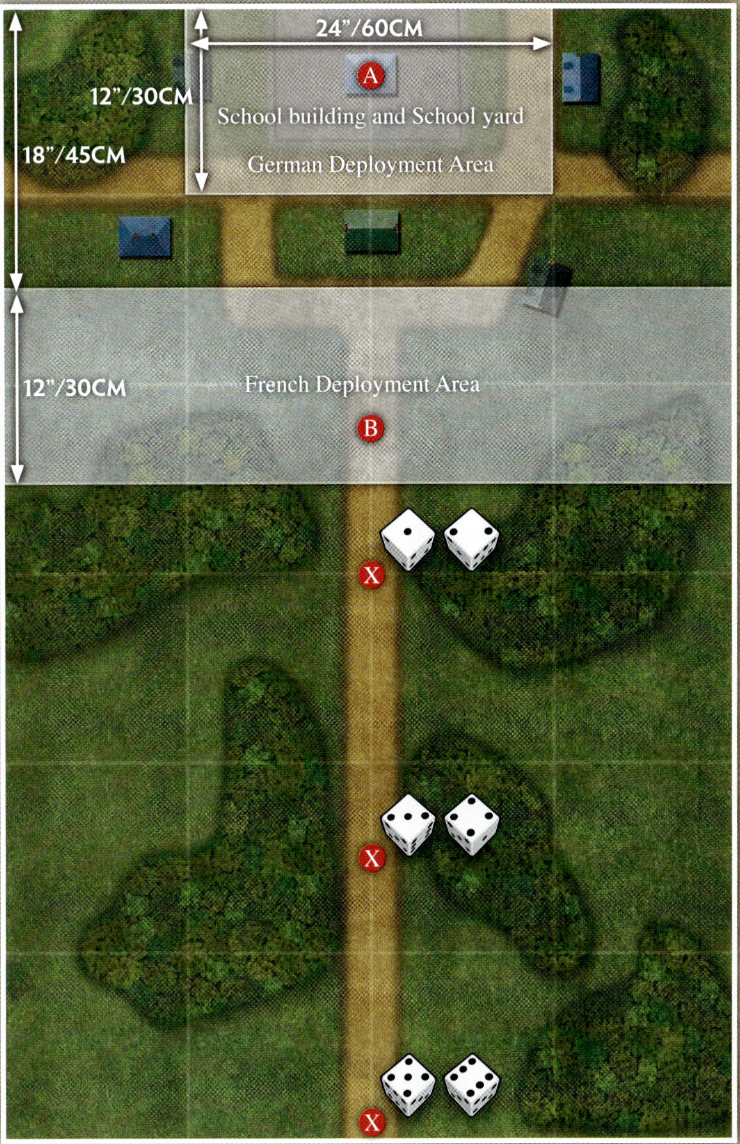

German Platoons must escape off this table edge.

YOUR ORDERS

GERMAN

Your comrades are being held in a small school building in Brasprats. You have reached them and as you begin loading them into the trucks, the alarm is sounded. Now you've got to get them back to friendly lines!

FRENCH

You have captured a large number of German paratroopers. As you decide what to do with them a convoy of American vehicles arrive apparently with orders to take the prisoners. Turns out its a ruse and the Germans are trying to escape with the prisoners. You must stop them before they can escape!

PREPARING FOR BATTLE

1. Set up the terrain as shown. Place the prisoner of war Fallschirmjäger Platoon inside the School building or School yard found in the area marked A.

2. Set up three Reserve Checkpoints at each point marked X.

3. The French player now deploys an FFI Company and their independent teams in the area defined by the scenario map marked B. The remaining two FFI Companies are held off the table and placed in Reserve.

4. The German player now deploys their entire force in the area defined by the scenario map marked A. The Fallschirmjäger Platoon that is not the prisoner of war platoon may begin the game mounted in the Captured Truck Platoon.

BEGINNING THE BATTLE

The German force has the first turn.

ENDING THE BATTLE

The battle ends:

- At the start of the French turn six, or
- When there are no German platoons left on the table.

DECIDING WHO WON

The German player wins if they manage to escape off the German table edge with the prisoner of war Fallschirmjäger Platoon. See the next page to determine how successful your raid was. Otherwise, the French player wins, having stopped the raid cold and inflicted an embarrassing defeat on one of Germany's elite units.

The imprisoned Fallschirmjäger are relieved to see thier comrades and rush the gates.

The alarm is set off and the FFI springs into action to halt Lepkowski's column.

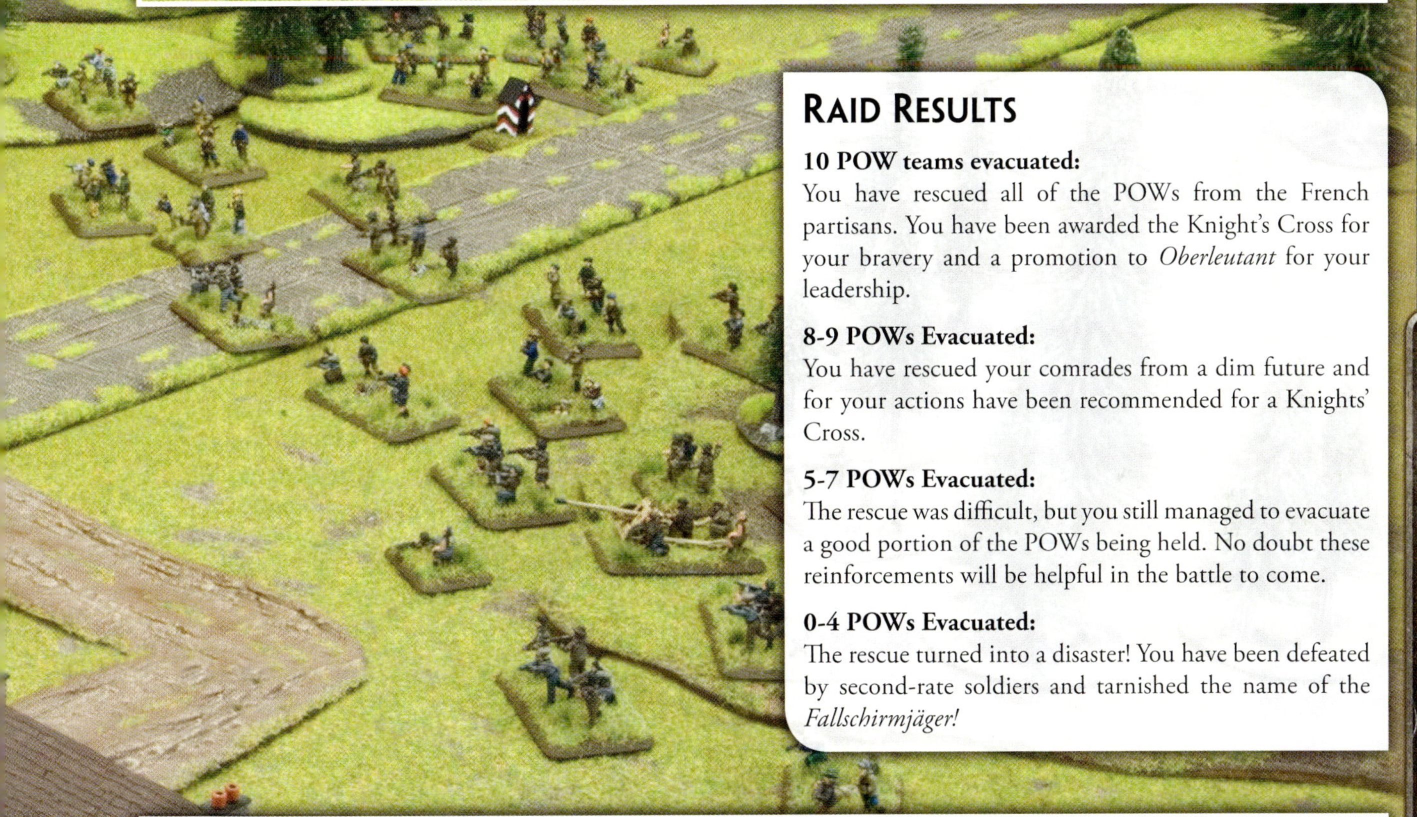

Raid Results

10 POW teams evacuated:
You have rescued all of the POWs from the French partisans. You have been awarded the Knight's Cross for your bravery and a promotion to *Oberleutant* for your leadership.

8-9 POWs Evacuated:
You have rescued your comrades from a dim future and for your actions have been recommended for a Knights' Cross.

5-7 POWs Evacuated:
The rescue was difficult, but you still managed to evacuate a good portion of the POWs being held. No doubt these reinforcements will be helpful in the battle to come.

0-4 POWs Evacuated:
The rescue turned into a disaster! You have been defeated by second-rate soldiers and tarnished the name of the *Fallschirmjäger!*

Lepkowski smashes through the FFI outposts and speeds away toward friendly lines and safety.

THE BATTLE OF THE LA FIÈRE CAUSEWAY
6-9 JUNE 1944

On D Minus One the men of the US 82nd 'All American' Airborne Division jumped into Normandy with the mission of halting German reinforcements from reaching Utah Beach. The division's two main drop zones were positioned around the French town of Sainte Mere Eglise in two large open areas. The Merderet River divided these two areas and thus became the first objective of the 82nd in order to consolidate their position.

During the drop, the shaken pilots of the paratrooper's C-47 transport planes dropped the soldiers all across the Cotentin Penninsula, resulting in mass confusion. Still, the men knew what had to be done, having studied the sand tables and maps thoroughly back in England.

Elements of the 507th Parachute Infantry Regiment converged on the bridge during the night of 5/6 June and established a rough perimeter.

At 1400 hours on 6 June, the Germans attacked in force with *1057. Grenadierregiment* supported by *100. Panzer Ausbildungs und Ersatz Abteilung*, a tank training and replacement battalion.

The grenadiers advanced with the old French training tanks. As they approached the bridge, the tanks stopped in front of the small minefield deployed by the paratroopers. As the infantry moved to clear the mines, the paratroopers opened fire, causing casualties and forcing the tanks to pull back.

Once they regrouped, a handful of H-39 tanks advanced slowly toward the bridge. Suddenly, the first was knocked out by two carefully hidden bazooka teams. Nevertheless the Germans pressed the attack and forced the paratroopers back to the eastern side of the bridge.

The Germans halted at the river, unable to advance through the now heavily-defended western bank. A stalemate developed as the two sides struggled to gather enough strength to push the other back.

Eventually the paratroopers managed to push the Germans back thanks to the timely arrival of their glider-borne regiment as well as several Sherman tanks which had arrived from Utah Beach.

Unable to repel firepower of that magnitude, the *Beutepanzer* tanks withdrew and the bridge finally fell into the hands of the Americans on 9 June.

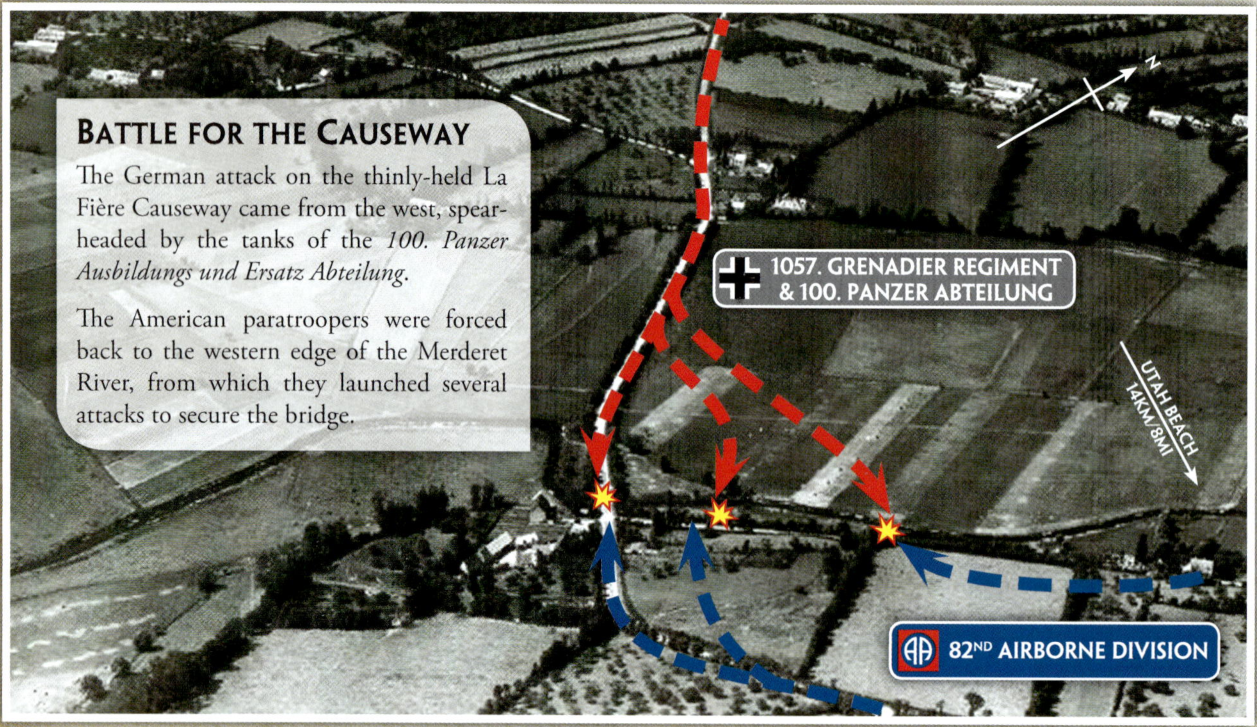

BATTLE FOR THE CAUSEWAY

The German attack on the thinly-held La Fière Causeway came from the west, spearheaded by the tanks of the *100. Panzer Ausbildungs und Ersatz Abteilung.*

The American paratroopers were forced back to the western edge of the Merderet River, from which they launched several attacks to secure the bridge.

SCENARIO FORCES

GERMANS

BEUTEPANZERKOMPANIE (CONFIDENT TRAINED)

Company HQ:
2x Panzer 35-R 731(f) tanks

Beutepanzer Platoon:
1x Command Panzer 35-S 739(f) tank, and
4x Panzer 35-R 731(f) tanks

Beutepanzer Platoon:
1x Command Panzer 35-S 739(f) tank, and
4x Panzer 35-R 731(f) tanks

Beutepanzer Platoon:
1x Command Panzer B-2 704(f) tank, and
4x Panzer 38-H 735(f) tanks

Beutepanzer Platoon:
1x Command Panzer IIIJ tank, and
4x Panzer 38-H 735(f) tanks

Bodenstandig Anti-tank Gun Platoon:
1x Command SMG team, and
3x 7.5cm PaK40 anti-tank guns

Bodenstandig Tank-hunter Platoon:
3x Marder I tank hunters

Bodenstandig Grenadier Platoon:
1x Command Panzerfaust SMG team, and
6x Rifle/MG teams

Bodenstandig Grenadier Platoon:
1x Command Panzerfaust SMG team, and
6x Rifle/MG teams

Bodenstandig Heavy Artillery Battery:
1x Command SMG team
1x Staff team
2x Observer Rifle teams, and
4x 15.5cm sFH414(f) Howitzers

Bodenstandig AA Gun Platoon:
1x Command SMG team,
3x 3.7cm FlaK43 anti-aircraft guns, and
3x 3-ton trucks

Luftwaffe AA Assault Platoon:
1x Command SMG team, and
2x 8.8cm FlaK36 with extra crew

AMERICANS

US PARACHUTE RIFLE COMPANY (FEARLESS VETERAN)

Parachute Rifle Company HQ:
Company Command SMG team
2iC Command SMG team, and
3x Bazooka teams

Parachute Rifle Platoon:
1x Command Rifle/MG team,
9x Rifle/MG teams,
1x 60mm mortar team, and
1x Bazooka team

Parachute Rifle Platoon:
1x Command Rifle/MG team,
9x Rifle/MG teams,
1x 60mm mortar team, and
1x Bazooka team

Parachute Machine-gun Platoon:
1x Command Carbine team, and
4x M1919 LMG teams

Airborne Engineer Combat Platoon:
1x Command Pioneer Rifle team,
2x M1919 LMG teams, and
2x Pioneer Rifle teams

Glider Rifle Platoon:
1x Command Rifle team,
9x Rifle/MG teams, and
1x Bazooka team

Glider Field Artillery Battery:
1x Command Carbine team,
1x Staff team,
1x Observer Rifle team, and
6x M3 105mm light howitzers

Glider Anti-tank Platoon:
1x Command Carbine team,
3x 57mm Anti-tank guns
3x Bazooka teams, and
3x Jeeps

The forces used in this scenario can be found in *Turning Tide* (US Parachute Rifle Company, page 144) and *Earth & Steel* (Beutepanzerkompanie, page 178).

Panzer B-2 740(f)
Char B1 bis
Code: FR070

Panzer 35-S 739(f)
SOMUA S-35
Code: FR050

Panzer 38-H 735(f)
Hotchkiss H-35/H-39
Code: FR020

Panzer 35-R 731(f)
Renault R-35
Code: FR030

THE BATTLE OF THE LA FIÈRE CAUSEWAY

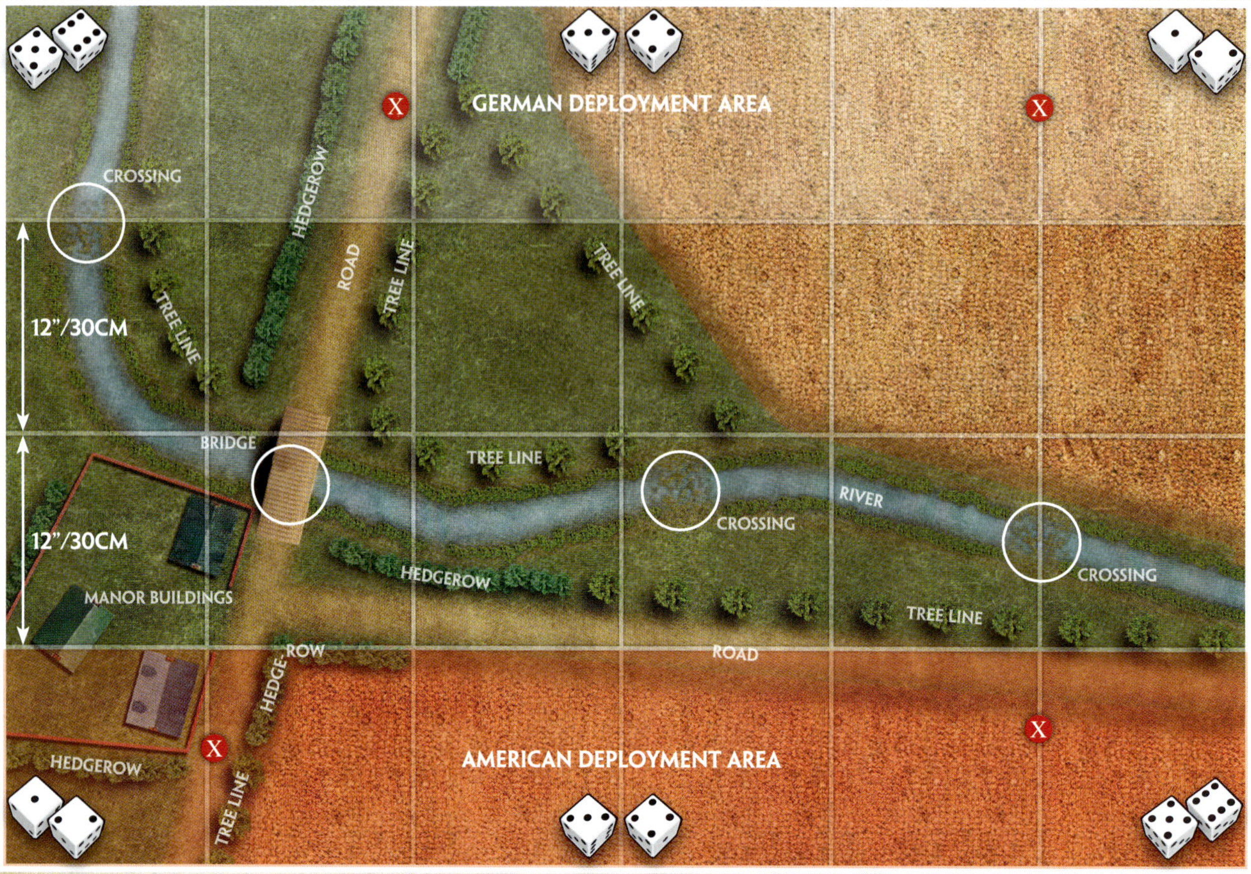

The La Fière Causeway is critical to both sides and must be captured at all costs.

This scenario uses the **Delayed Reserves, Meeting Engagement,** and **Scattered Reserves** special rules (see pages 264 to 271 of the rulebook).

Your Orders

German

Paratroopers have landed astride the Merderet River. Furthermore a large Allied assault force has landed due east of your position. Your mission is to secure the causeway and keep the way open for our reserves.

American

The causeway links the 82nd Airborne's drop zones. You need to secure this crossing so that we can gather our strength and make sure no Germans can reinforce Utah Beach.

Preparing for Battle

1. Set up the table as shown on the scenario map.

2. Place an objective at each point marked with X

3. Starting with the German player, both players now nominate at least half of their platoons to be held off the table in Delayed and Scattered Reserve at the start of the game.

4. Both players, starting with the German player, now alternate placing remaining platoons in their own deployment areas. The platoons may not be placed within 12"/30cm of the centre line of the table.

5. Again, starting with the German player, both players now place their Independent teams in their own deployment areas.

Beginning the Battle

1. Both players roll a die. The player with the highest result has the first turn.

2. Remember that this mission uses the Meeting Engagement special rule so, for the first turn of the game, all teams are considered moving at the beginning of the battle.

Ending the Battle

The battle ends when either:

- a player starts their turn in possession of any of the objectives in the enemy deployment area.

Deciding Who Won

The player that took their objective wins the battle. They have won control of the La Fière Causeway, forcing the enemy to have to retreat.

Calculate your Victory Points using the Victory Points Table on page 275 of the rulebook.

If neither side won use the There Are No Draws special rule (see page 275 of the rulebook) to determine Victory Points.

TERRAIN

The Battle of the La Fière Causeway scenario includes rivers, bocage, tree lines, fields, roads and more. Here are some suggestions for how to define a few of the major pieces of terrain for this scenario.

The Merderet River

The main feature of the battlefield is the Merderet River which runs across the table. In this scenario the river is rated Very Difficult Going. The crossing points are rated Difficult Going. The bridge is rated as Cross-country

The banks of the river have a lot of low-lying vegetation allowing teams, taking cover behind it, Concealment.

In June 1944 the river was flooded by the Germans making it a mire of mud and water. For our scenario we ignored this as it would bog down the game. However, if you would like to represent this make all ground within 6"/15cm of the riverbank Difficult Going.

Hedgerows

The hedgerows in this scenario have not grown large enough to be rated as bocage. Instead they are rated Very Difficult Going. They are tall enough and thick enough to block line of sight through them.

Tree Lines

The remainder of the tall vegetation are thin tree lines. These trees are not thick enough to block line of sight, but provide Concealment for teams on either side of them being shot at from the other side.

OTHER PLACES, OTHER TIMES

While the operation was a success for the US paratroopers, how would they have fared if there happened to be a better equipped German force in the area? Try replacing the Beutepanzerkompanie with a 1500 point Panzerkompanie from page 52 of *Earth & Steel*, or an SS-Panzerkompanie from page 150.

If you want to keep to the captured equiment theme, you could refight the battle over the river Orne, or Canal de Caen. Rather than US Paratroopers, use a 1500 point British Para force from page 20 of *Turning Tide*, trying to halt a counterattack against a Beute StuG Battery.

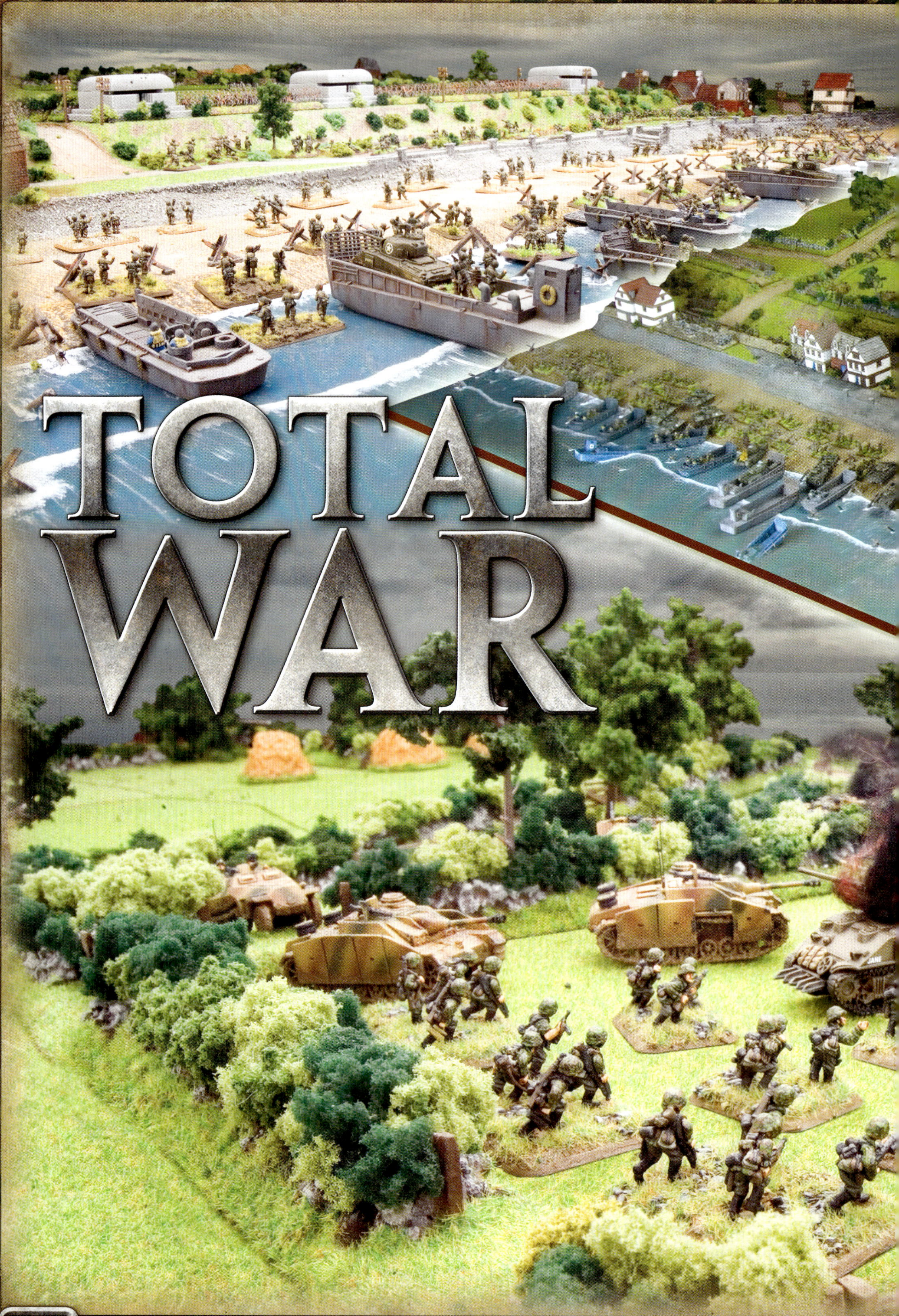

TOTAL WAR

POLLY
SUZY

TOTAL WAR

There comes a point in every gamer's life when they want to do something big. Total War is designed to do just that—put big armies on big tables. Instead of a mere company, you can field your entire collection at once.

Just think, thousands of points of *Flames Of War* troops arrayed across a large table, multiple companies under your command. The infantry holding the line on the right while the tanks advance forward on the left. The ebb and flow of battle begins, your tanks rolling up the enemy force as it advances on the left, your gallant infantry first holding the line then surging forward on the right.

With the battlegroup rules you have the chance to put any size force on the table and go head to head with someone else. There is no limit to the points you can field, so long as the table is large enough. And how much better are the bragging rights once you have beaten your enemy on a 20 foot/six metre-long table?

Total War also gives you a chance to change the flavour of your games, get a bunch of friends together, and play a big multi-player game. Instead of facing your friends across the table, join forces with them to take on another team of friends. Order some pizza and drinks and make a real event of it!

What Is Total War?

Total War is a large point level game of *Flames Of War*, played by two or more players. In normal games you tend to play at the company level, with games averaging about 1500 points or so. However, in Total War you will fight alongside other companies, forming a battalion level game with armies on both sides worth 4000 points or more!

Why play Total War?

Total War is geared towards having fun. It is more of a social game than a competitive one. To that end, the Total War rules and missions are designed to be fun and intriguing. Though the forces are balanced against each other, Total War is not entirely focused on who are the winners and losers but rather on making sure all players have a fun and interesting game, while also giving players the excuse to do daring things, field odd forces, or play with a lot more points than they'd normally get to use on the battlefield.

PLANNING A TOTAL WAR GAME

Your tanks have smashed through the enemy's defences. Your friend's infantry company has forced its way through the town. He has cut off the only remaining enemy reinforcements with any chance of stopping your well-planned offensive. Teamwork has brought you victory!

Total War is all about trying something new and different. It's your chance to put more models on a table than you've ever done before. It's your chance to find out just how nasty a whole company of Tiger heavy tanks is, and to find out if they can stop a whole battalion-sized combined-arms task force or even an entire regiment of Stalin heavy tanks!

With *Total War* you can play bigger games with more friends than you have ever played before. In the end though, what *Total War* is really about is having even more fun!

PLANNING

Like most things, *Total War* battles run more smoothly if you do some planning beforehand. While you can set up a small *Total War* game on the fly, the size of the forces and the number of players involved in a larger game requires a bit of organising ahead of time.

The most important thing to organise is a place to play. Since *Total War* games usually take longer than the usual game, you may need to book a table. Make sure you inform the store owner or club organiser how long you need the table for. If you are playing a really big game, you may need to book a hall or get a store owner to put all of their tables together for a big event.

Once you have arranged a table, make sure that you tell all of your players when the game is and what they are expected to bring in the way of terrain and forces.

SETTING UP A TOTAL WAR GAME

Pick terrain which is compatible with the type of armies you are using. Armoured forces need open terrain to manoeuvre while infantry can fight better in rough terrain and close quarters. Lots of forests and towns with large armour forces will slow down your date with destiny considerably.

Before you consider the table-size, the type of terrain, and terrain set-up you need to take into account the army size, playing area, mission, and time allowed in playing the game.

TABLE SIZE

Large tables offer a great place to fight a large battle. Picking the table to match the battle will give your battle flavour, multiple tactical options, and more realism.

Don't forget to provide space off the table to move around and a place to store your armies before they enter the battle. We've included some helpful tips on choosing the right table size for your game on page 105.

TYPE OF TERRAIN

With larger tables, it is important to consider how much terrain you use and how it is laid out. You might need to pool terrain together to fill out a large playing area. Bigger tables and bigger forces also mean bigger terrain features.

A forest may need to be big enough to engulf an entire company while villages should be large enough to force movement through them instead of around them. Roads and rivers become dominant terrain features so be careful in their placement. You can cut off an entire board section with an ill-placed river giving one side a safe area behind the river and forcing movement away from the river into a smaller area of the battlefield.

Matching your terrain to your battle and its size is important. Ensuring players are afforded the opportunity to bring the right forces for your terrain board will increase the fun of fighting a large battle.

TERRAIN SET-UP

Group several woods together to make a larger wooded area and cluster houses to form one or more villages. This allows more scope for whole companies to get involved in a fight in the woods or in clearing a village generating more drama in the game. Remember though that big woods and big villages bring big fights. If that is what you want then go for it.

Combining terrain is also beneficial. Placing a village next to woods will enhance the battlefield. It will provide a major terrain feature to confront the tactical guile of your players.

Make sure your terrain facilitates the battle you want to play. If you want to use rangers have obstacles for them to conquer. Paratroops need a landing area they can hold.

Simulating real world terrain is also exciting. Aerial photographs or military maps can help you set up historical battlefields. Hills, roads, woods, and rivers become even more meaningful when they are real.

Finally, never leave a big void in the centre of the table, as it will become a deadly killing field that will stifle manoeuvre.

MUSTERING A TOTAL WAR FORCE

PICKING FORCES

You need to make sure that your players know what type of army they need to build and bring to your battle and to give them time to put it together. Team games should range from 2500 to 4000 points for two player teams to 3000 to 6000 points for three or more player games. If you go beyond 6000 points, you may need a bigger table and more terrain (not to mention more time!).

There are two ways of organising this. You can either give each player a points total to build their force and then combine them on the day, or can have the players get together to create a single integrated force in which each player's company has a designated role.

CHOOSING A THEME

While you can just tell everyone to bring along a force, it can be much more fun to choose a theme for the game. This could be a historical battle or an interesting tactical situation that you have created a scenario for.

You can take a theme as far as you like. You could just theme the terrain and the forces to generally match a particular battle, or you could go the whole hog and create a couple of centrepiece terrain features with some scenario special rules and do some research on what actually fought at your battle. Your theme could be as simple as 'You take a tank and an armoured infantry company, and we will take some paratroopers with some tanks and infantry coming to the rescue.'

Regardless of how far you choose to take your theme, spend a few minutes discussing the game and set a few guidelines. That way both sides can have specific objectives to accomplish and can build their armies based on how they believe they can win the battle.

TRY THINGS OUT

Total War is also a great way to try out units you wouldn't normally take. Units like a full self-propelled artillery battery or a full platoon of *Elefant* heavy tank-hunters that cost lots of points can be a bit expensive for smaller forces, yet they fit nicely into a large *Total War* game. Taking plenty of big nasties like these make your force really hard. Of course, they also chew up points and can leave you open to being outmanoeuvred on a larger table. This makes for an interesting challenge as you try to balance your unstoppable punch with just enough troops to hold their flanks!

Alternatively, you can build a Support Company list (see page 98). This gives you the excuse to field weird and wonderful forces that you wouldn't normally see on the *Flames Of War* battlefield.

Even low-points units change in bigger battles. Hordes of medium tanks get really big and massed infantry formations positively enormous! Unfortunately, the enemy has even more weapons to chew them up, so you'll need some really tough support to cover their advance too. Think about how much time you have before you create too big a horde, it can take a while just to move all of your models!

Other equipment like trucks and half-tracks have a relatively limited role on a smaller table, but can prove invaluable for moving infantry and guns into position on a big table. Likewise, long-range anti-tank guns and artillery come into their own when there is enough space for the enemy to avoid their shorter-ranged cousins.

MULTI-PLAYER GAMES

Total War is ideal for big multi-player games at the club or in your local store. Have everyone bring a force, then divide players into two teams with roughly the same number of players and the same number of points.

Big multi-player games are great when you have lots of players and not enough tables. Push a couple of tables together and you can have a awesome game with six or eight players without taking up much more space than a normal two-player game.

If you can, let everyone know when the game is and pick their teams in advance.

TEAM GAMES

In a team game each player commands their own force. Although all of the players' Support Platoons are pooled in a single Support Group, each player will normally command their own Support Platoons. If they want to, they can loan their Support Platoons to other players, or even mass them under a single commander.

All players on the same team take their turns simultaneously, completing their Starting, Movement, Shooting and Assault Steps together at the same time.

Any special rules that apply to a player's force only apply to their own troops. Any Warriors that they field can only join that player's own platoons and only affect their own troops.

COORDINATED BATTALIONS

When building your forces for a Total War game, your team may want to try and configure your force for a specific objective, such as speed, defence, etc. While building your lists, discuss with your teammates if you'd like to field a specific type of Battalion. If your team meets the requirements as one of the Battalions listed below, then your force will gain that rule during the Total War game. Each team may only choose one type of Battalion to use in the game, even if it qualifies for several types.

Tank Battalion

If at least half of the companies in your team are Tank Companies, it is rated as a Tank Battalion and gains the following rule.

Keep those tanks moving: *All tank teams in your team may re-roll failed attempts to free Bogged Down vehicles and failed attempts to remount Bailed Out vehicles.*

Infantry Battalion

If at least half of the companies in your team are Infantry Companies, it is rated as an Infantry Battalion and gains the following rule.

Needed on the front line: *Platoons entirely made up of Infantry and Man-packed Gun teams may move At the Double through Difficult Going.*

Mechanised Battalion

If at least half of the companies in your team are Mechanised Companies, it is rated as a Mechanised Battalion and gains the following rule.

In and out: *During the Assault Step, you may dismount any Infantry or Gun teams. Teams that dismount this way cannot launch an assault in the turn they dismount.*

Artillery Battalion

If at least half of the companies in your team are Support Companies with Artillery Platoons as their Combat Platoons, it is rated as an Artillery Battalion and gains the following rule.

Forward Direction Centre: *All players' Company Commanders and Observer teams may act as Spotting teams for another player's Artillery platoon, even if they are Allied.*

Air Support Battalion

If at least half of the companies in your team have Air Support, it is rated as an Air Support Battalion and gains the following rule.

Air Operations: *Each Player may keep their Air Support Pools separate instead of choosing the best level as normal (see page 178 of the rulebook).*

Combined Arms Battalion

If your team has at least two different types of companies (Tank, Mechanised, Infantry, Support, or Fortified) it is rated as a Combined Arms Battalion and gains the following rule.

Calculated Risk: *Every player may choose up to one Platoon in their force to gain Spearhead Deployment (see page 261 of the rulebook) at the start of the game.*

TOTAL WAR SPECIAL RULES

Total War games bring multiple companies together. These can be from different nations, forces with similar support options, and even forces led by famous leaders. The following special rules cover these circumstances, which are important to remember when putting together your Total War force.

INTER-ALLIED COOPERATION

Most armies, despite the best will in the world, find it impossible to cooperate as closely with their allies as they would like. Different languages and doctrines combine with incompatible equipment and procedures to render close coordination impossible.

Companies and platoons from the armies of different countries are Allied companies and platoons (see the Allied rules on pages 68, 70, and 105 of the rulebook).

WARRIORS

Where would an all-out war be without heroes?

In Total War you can field as many Warriors that you want, as long as there is only one of each named Warrior and they are fielded with the company that they are normally fielded with.

AIR SUPPORT

Air support is allocated to a mission, not to a particular unit fighting to fulfil that mission. This allows this highly mobile resource to be used in the most effective place on the battlefield.

Remember, a force (or a side in Total War) only ever has one Air Support Pool shared across all of its companies. If multiple companies have Air Support, take the best level as the Air Support level for the entire force, unless the force contains an Air Support Battalion (see page 96).

FORCE MORALE

With all of the companies running around in a single force, it's good to know where the breaking point is for your force.

Unlike normal, a Total War game does not necessarily end when a company fails its Company Morale Test. Instead, each Total War mission will define exactly how your troops handle failed Company Morale Tests.

FIELDING A SUPPORT COMPANY

Have you ever wanted to field a Marder tank-hunter company, or perhaps you would like to command a whole artillery battalion? Well here's your chance!

In Total War games some players may want to try fielding a Support Company using troops that normally can only be fielded as support platoon. To do this use the following instructions:

Step 1: Choose a Company Diagram

Choose any Company Diagram from the nation you intend to play. You may want to chose one that matches the theme of the Total War game you are playing, or perhaps you might want to choose the same one being used by one of your allies.

Step 2: Choose your Combat Platoons

Chose one of the Support Platoons from the Company Diagram you have selected in Step 1. This platoon will determine what type of company your Support Company will be. The selected Support Platoon becomes your force's Combat Platoons. You must take two or three Combat Platoons. You may ignore any special rules that require you to take other platoons before the platoons you have selected as your Combat Platoon.

If your Combat Platoon has multiple variants (such as different Motivation or Skill ratings), you must chose one and stay with it across all of your Combat Platoons.

For example, if you choose a Field Artillery Battery, your company will be an artillery Support Company and you must field at least two but no more than three Field Artillery Batteries.

Step 3: Choose your Support Platoons

All of the other platoons on the Company Diagram are your Support Platoon options. You may only select one platoon from those that would normally be Combat Platoons, and only one platoon from those that would normally be Weapons Platoons, as Support Platoons. You may also choose one platoon from each Support box, including Support Platoons of the type that you chose to be your Combat Platoons, as further Support Platoons.

Step 4: Choose your Headquarters

If all of the teams in your Combat Platoons are tank teams, you do not have a Company HQ platoon. Instead, before deployment, you may remove one tank from any of your Combat Platoons and make it into the Company Command tank.

If your Combat Platoons include mixed types of teams, use the directions on page 99 to make up your Support Company Headquarters. You must chose the variant that has the same Skill Rating as the Combat Platoon you chose in Step 2. The teams in your headquarters retain all of your nation's usual special rules.

Support Armies in Normal Games

Support Companies are not the most balanced forces out there and should not be used in tournament play. However, with the permission of your opponent, you can certainly play a Support Company outside Total War in casual *Flames Of War* games using the normal missions.

HEADQUARTERS

COMPANY HQ

Headquarters	Veteran	Trained	Conscript
Fearless Company HQ	40 points	30 points	25 points
Confident Company HQ	35 points	25 points	20 points
Reluctant Company HQ	25 points	20 points	15 points

OPTIONS

- Add two Jeeps for +5 points.
- Add two Half-tracks for +10 points.

FIELDING YOUR SUPPORT COMPANY

The characteristics for the Rifle teams, jeeps, and half-tracks are given below. These are meant to be simple generic teams and are not nation-specific. Feel free to use whatever models you would like for these teams, but they will use the following Arsenal entries. The teams will still use all of their nation's special rules, such as Automatic Rifles for the Americans and Stormtroopers for the Germans.

ARSENAL

If you'd like to field transports that match your nation but are unsure which models to use, here are some suggestions.

For Germans use Kübelwagen models for your jeeps and Sd Kfz 251 models for your half-tracks.

For Americans, British, and Soviet forces, use the standard American-made Jeep transport for your jeeps and either the M2 (for British and Soviet) or the M3 (American) as your half-tracks.

For other nations, have a look at gun platoons from their briefings to get an idea what sorts of transports you should use (if any).

INFANTRY TEAMS

Team	Range	ROF	Anti-tank	Firepower	Notes
Rifle team	16"/40cm	1	2	6	

TRANSPORT TEAMS

Vehicle	Mobility	Armour Front	Armour Side	Armour Top	Equipment and Notes
Jeep	Jeep	-	-	-	Passenger-fired AA MG.
Half-track	Half-tracked	1	0	0	Passenger-fired AA MG.

SUPPORT ARMY EXAMPLE

In this example, I have created a 1650-point Support Company with M7 Priest HMC self-propelled howitzers.

Step 1: Choose a Company Diagram

The company has been formed from the US Rifle Company found in *Devil's Charge* on page 48.

Step 2: Choose your Combat Platoons

I would like to build my force around the cool M7 Priest HMC self-propelled howitzers, so I chose the Armored Field Artillery Battery as my Combat Platoon.

I will need to take two or three Armored Field Artillery Batteries.

Since my platoon has multiple ratings (Confident Veteran and Confident Trained) I will need to choose one and stick with it across all of my Combat Platoons. In this case I have chosen Veteran artillery.

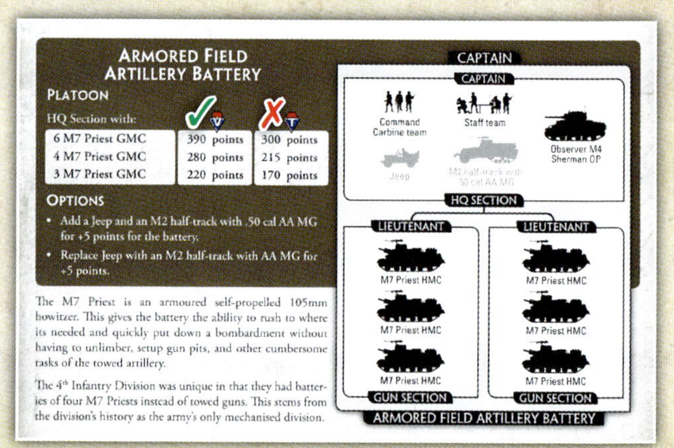

ARMORED FIELD ARTILLERY BATTERY

PLATOON

HQ Section with:	✓ V	✗ T
6 M7 Priest GMC	390 points	300 points
4 M7 Priest GMC	280 points	215 points
3 M7 Priest GMC	220 points	170 points

OPTIONS

- Add a Jeep and an M2 half-track with .50 cal AA MG for +5 points for the battery.
- Replace Jeep with an M2 half-track with AA MG for +5 points.

The M7 Priest is an armoured self-propelled 105mm howitzer. This gives the battery the ability to rush to where its needed and quickly put down a bombardment without having to unlimber, setup gun pits, and other cumbersome tasks of the towed artillery.

The 4th Infantry Division was unique in that they had batteries of four M7 Priests instead of towed guns. This stems from the division's history as the army's only mechanised division.

Step 3: Choose your Support Platoons

Having chose the Rifle Company, I have a lot of different platoons to take as support. I'd like to keep with the artillery theme, so I'll grab something from one of the Artillery Boxes.

I still have to obey the normal rules that say I can only chose one platoon from each box, so I'll have to choose between a Calliope Tank Platoon, a Chemical Mortar Platoon, a Field Artillery Battery, or a Rocket Launcher Battery. I like the Calliope, so I'll choose that platoon from this box.

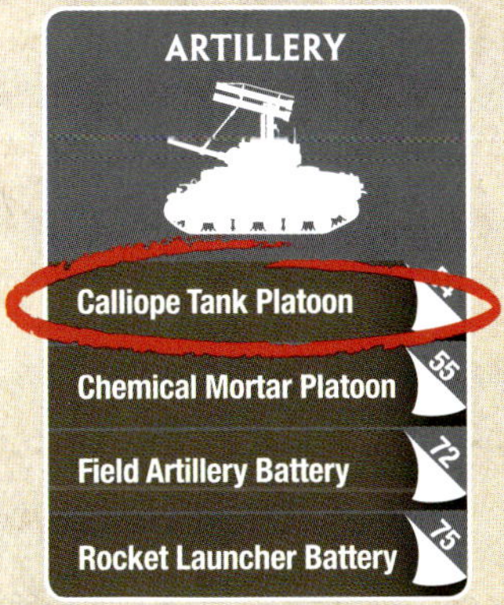

I can also take a platoon from the box I chose in Step 2, so I'll grab a veteran Field Artillery Battery (155mm Gun SP) for some heavy firepower. Obviously I'll need the battalion's L4 spotter plane, so I've taken an Air Observation Post as well.

To round things out I'd like to have some troops in the field to help me spot targets and help out where needed. So I have added a Cavalry Recon Platoon with three Cavalry Recon Patrols.

Step 4: Choose your Headquarters

Not all of the teams in my Armored Field Artillery Batteries are Tank Teams, so I'll use the Support Company Headquarters platoon. My platoons are Veteran, so I'll need to use the matching set of points for my HQ (35 points).

Since my combat platoons are mobile, I should probably give my HQ some wheels. I've added half-tracks to my Support Headquarters for +10 points. My force is American, so I can use standard M3 Half-track models, or I could use a captured German one as a bit of flavour!

After all of the dust has settled, I've got an impressive collection of firepower to help out my comrades:

My Armored Field Artillery Battalion

Support Company HQ:
At full strength with half-tracks — 45 points

Armored Field Artillery Battery:
At full strength with half-tracks added — 400 points

Armored Field Artillery Battery:
At full strength with half-tracks added — 400 points

Armored Field Artillery Battery (155):
At full strength — 335 points

Callopie Tank Platoon:
4x Calliope Tanks — 160 points

Cavalry Recon Platoon:
At full strength — 270 points

Air Observation Post:
L4 Grasshopper AOP — 40 points

Total 1650 points

TOTAL VICTORY MISSION

While you can play *Total War* with the normal missions from the rulebook, the sheer size of the game demands a bigger mission. The normal missions end with the capture of the first objective. With the bigger forces and tables of *Total War*, there needs to be more objectives and the ability to counterattack to recapture lost objectives.

The Total Victory mission works quite differently from normal *Flames Of War* missions. Instead of winning the game by capturing an objective in this mission, you win by taking and holding objectives throughout the game. Each turn you gain a victory point for each objective that you hold, making it important to secure objectives early and to hold them against enemy counterattacks while attempting to contest the enemy's objectives.

This requires a change in the way the game ends. In most *Flames Of War* games the game ends with the capture of the objective or the destruction of one company or the other. In Total Victory, players fight for a set period of time. When choosing how long to fight, you need to allow enough time to reach some sort of conclusion, but not so long that the game draws out into a hunt for the last few surviving enemy.

We find setting a strict time limit on deployment to be useful for large games as it reduces the amount of time wasted before the game begins. It focuses players on getting their troops deployed and ready for battle rather than chatting, eating, and all those other distractions.

You'll also notice that the Total Victory mission allows you to bring reconnaissance, airborne, and commando units on to the table from behind your opponent! This represents them sneaking through or landing behind enemy lines to launch a raid. Not only does this add a new dimension to the game, but it gives these types of troops a whole new role in the game. They certainly make an all-round defence of objectives and artillery positions a good thing to consider!

Finally, you don't just have to play the Total Victory mission plain. Spice it up by adding some historical background to the mission. For instance, you might choose to place the terrain and the objectives to match a particular battle, or to deliberately give one side a bigger force.

TABLE SIZES

With huge forces, you'll find a bigger table opens up new challenges and adds a new dimension to your *Total War* games. The suggestions below give a guideline for tables sizes for larger games. Don't worry if you don't have a table this big, just play on what you have.

It's important not to make the table too wide for players to reach the centre, so keep the width to a maximum of 6'/1.8m. Rearrange the tables into an L, O, T, U, or Z shape to create even more interesting games.

1500 to 5000 points 8'/2.4m x 4'/1.2m to 6'/1.8m

One-on-one *Total War* games will fit nicely on a double-sized table. Offsetting the tables in an L, S, or T shape makes for interesting challenges and makes reaching the centre easier.

Up to 7500 points 12'/3.6m x 6'/1.8m

Multi-player *Total War* games need a bigger table. Depending on the number of units, you may need an even bigger table.

Up to 10,000 points 16'/4.8m x 6'/1.8m

A massive game this size will need a really big table. A table this size makes long-range artillery and trucks to move your infantry vital.

Global War 24'/7.2m x 6'/1.8m

This is the upper limit of practical table size. Any bigger than this and the game will break into several smaller battles.

TOTAL VICTORY SPECIAL RULES

TOTAL WAR RESERVES

Even though the forces used in *Total War* games are large, the normal Reserves rules work fine. Although only a few reserves will arrive in the first few turns, these larger forces are big enough to last until the reserves start flooding in on later turns.

Where the reserves rules need to change for these games is to allow airborne and reconnaissance troops to outflank the enemy and attack them from the rear. This represents them infiltrating through difficult terrain and weak spots in the enemy lines (or simply parachuting in behind them) and seeking out vulnerable enemy headquarters and artillery.

This mission uses the normal Reserves rule on page 268 of the rulebook.

However, any Reconnaissance, Fallschirmjäger, Parachute, Airlanding, Glider, Commando, Ranger, or horsed cavalry platoon that arrives from Reserve may enter the table from either their own table edge, or from any other table edge where their entry point is at least 8"/20cm from all enemy teams except Independent and Warrior teams.

BATTLEGROUP MORALE

With the large forces used in *Total War*, battles do not end with the destruction of the first company to die. Instead, forces fight down to the last company.

While the knowledge that other companies are there to help out keeps a company going when things get rough, being the only troops left is equally dispiriting to the last surviving company.

Unlike normal Flames Of War *games with multiple companies, a* Total War *game does not end when a Company fails a Company Morale Check.*

The game continues until one side starts their turn with at least half of their Companies Destroyed.

MODIFICATION SUGGESTIONS

When planning your Total War mission you can really go all out and do whatever you like. You can have airlanding operations flank the back of the table, while the rest of your assault troops push forward on foot. You could even have a game where your opponent must retrieve an objective and then escape off the table edge.

One interesting suggestion is to blend some of the cool scenario rules found in this book into one massive game. For example, you may want to launch an amphibious assault on one table using the Amphibious Assault rules on pages 20 to 25, while conducting an airborne landing on an adjacent table using the Airborne Assault rules on pages 12 to 17.

The possibilities are limited only by your imagination, so get with your group and write a truly epic Total War mission!

TOTAL VICTORY

Total Victory uses the **Prepared Positions** (see page 264 of the rulebook), **Battlegroup Morale**, and **Total War Reserves** (see page 103) special rules.

PREPARING FOR BATTLE

1. Decide on a start and ending time for the game. You will probably need half an hour, plus an extra half an hour for every 1000 points on a side. Make sure you have an extra hour available after the end time so that you can finish both players' turns.

2. Mark out the centre line on the table. This should either run the length of the table or across the table on a diagonal.

3. Both sides now roll a die. The side with the higher score chooses one side of the centre line to attack from. The other side defends from the opposite side.

4. Starting with the attackers, both sides place an objective on the attackers' side of the centre line at least 12"/30cm from the table edges and the centre line.

5. Starting with the defenders, both sides place an objective on the defenders' side of the centre line at least 12"/30cm from the table edges and the centre line.

6. Starting with the attackers, both sides place an objective within 6"/15cm of the centre line, but not within 12"/30cm of the table side edges.

7. The attacking side now has up to five minutes to plan their strategy and an additional 20 minutes to deploy their entire force including Independent teams, but excluding Reconnaissance Platoons.

   They may deploy anywhere on their side of the centre line as long as they are at least 12"/30cm from the centre line.

   All teams in a Combat Company must deploy within 32"/80cm of their Company Command team.

   The attackers may chose to keep any platoons they wish in Reserve. Any platoons not completely deployed within 20 minutes must be held in Reserve.

8. The defending side now deploys under the same conditions and restrictions, but must also deploy any Reconnaissance Platoons that are not held in Reserve at this time.

9. The attacking side now deploys any Reconnaissance Platoons that are not held in Reserve.

Reconnaissance Platoons may deploy as normal or they may deploy up to the centre line as long as they are at least 16"/40cm from all enemy teams.

BEGINNING THE BATTLE

1. Both sides start the game in Prepared Positions, so their troops are in Foxholes and Gone to Ground.

2. The attacking side has the first turn.

ENDING THE BATTLE

The battle ends as soon as both sides have had the same number of turns and the agreed upon ending time has passed.

If the game has reached a critical moment and both sides want to see what happens next, give each player another turn or two to find out.

DECIDING WHO WON

At the start of a side's turn when you normally check victory conditions, that side gains one Victory Point for each objective that it holds. An objective yields a Victory Point to the side that has taken it every turn that they hold it.

It is a good idea to keep a running total of Victory Points scored as you play the game. At the start of each turn add up the Victory Points you scored this turn, and add them to the running total.

A side starts the game holding all objectives in their deployment area. They hold any objectives that they have taken (even if all of their troops have since moved away) until the enemy takes it back or contests it by having troops that could take it within 4"/10cm.

The side that has the most Victory Points at the end of the game wins.

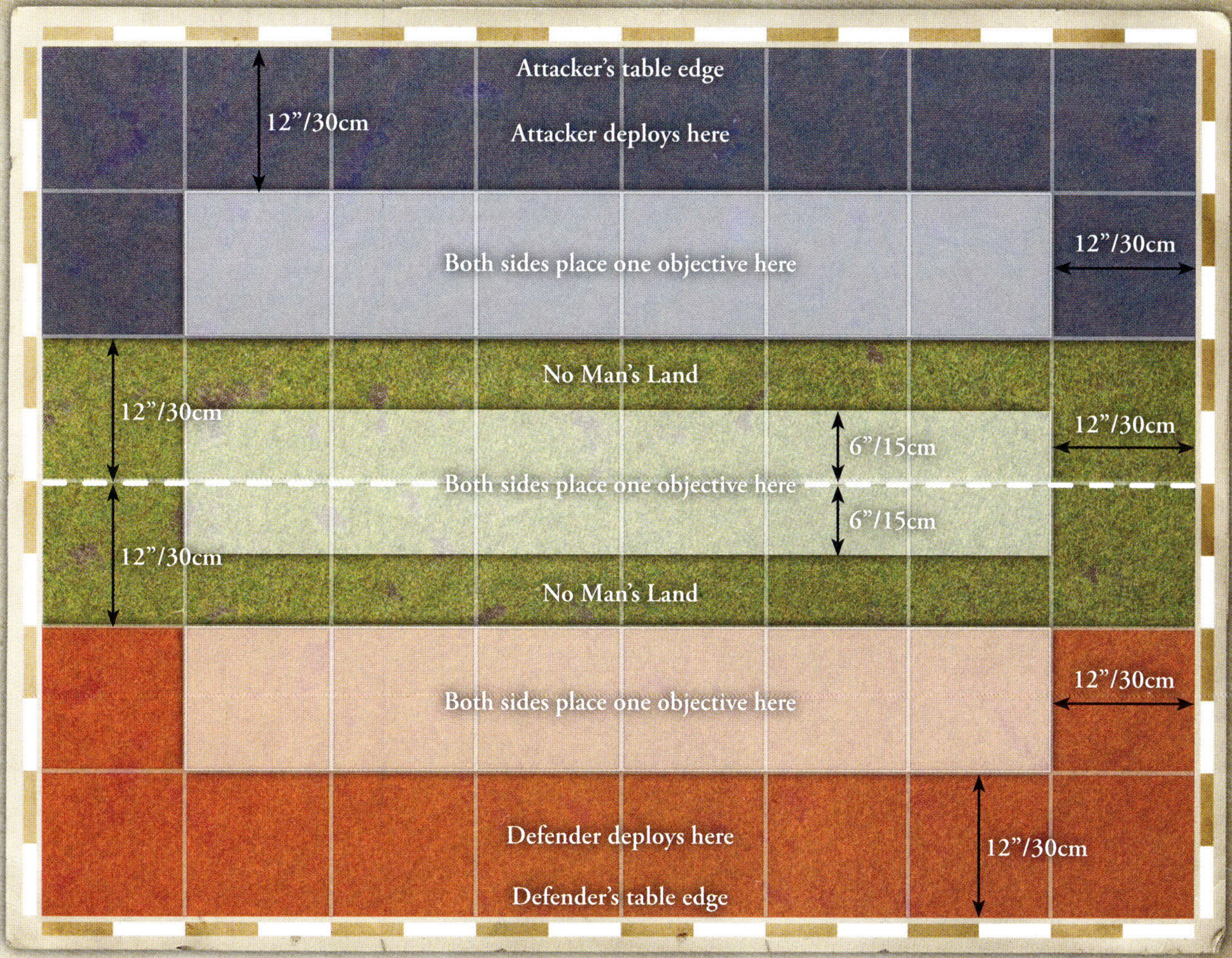

OTHER TABLE LAYOUTS

There are many other ways to lay out your table for the Total Victory mission. Try to avoid long and skinny configurations as these limit the frontage on which you can fight. On many table configurations a diagonal centre line works best and can produce interesting tactical situations as players attempt to capture objectives in two distinct parts of the battlefield.

NARROW TABLES

If you find yourself playing on a long narrow table, you could find that you don't have enough room to place the objectives in the players' deployment areas.

You can solve this in two ways: either run the centre line diagonally across the table, or reduce the minimum distances between the objectives and the table edges and centre line to 8"/20cm instead of 12"/30cm.

RIVER RUSH MISSION

During the campaign in Normandy, rivers played important roles. They were easier to defend than open ground by channelling the enemy's advance toward crossings and bridges. This makes for a strong defence with a small amount of troops. River assaults were hard-fought and intense, magnified even greater when an important prize, such as Caen or Paris lay on the opposite bank. The River Rush mission reflects such an important assault. In fact it is so important that both sides have amassed a huge number of troops in the area. This will be no simple fight!

THE RIVER

The River Rush mission is unique from others in that it requires a very specific piece of terrain: a river. All of the Crossing Rivers rules can be found in the rulebook on pages 56 to 57.

Two bridges and one natural ford cross the river. The three crossings should be spread out evenly across the length of the river. There should be no length along the river greater than 24"/60cm between crossings. If you are playing on a particularly deep table with a long river, you may need to add a few more bridges or fords.

The bridges have been rigged for demolition by the defenders, meaning the attackers will need to grab them quickly! The ford is a natural crossing point and therefore cannot be demolished.

FORTIFICATIONS

Since this is a hasty battle on both sides, no player may place any fortifications within 12"/30cm of the river.

RIVER RUSH SPECIAL RULES

Bridge Demolition

The best way to deny a crossing to the enemy is to utterly destroy it!

All of this mission's bridges have been prepared for demolition (but not the natural river crossing).

If at the beginning of their Movement Step, a player has one or more Infantry teams (other than Independent Teams) adjacent to a bridge that has been prepared for demolition, roll a die for each team adjacent to the bridge:

- *If the result is 6, the bridge is demolished and collapses. Any teams on the bridge are immediately Destroyed.*
- *Otherwise the bridge was not ready for demolition and nothing happens this turn.*

Pioneer teams may re-roll failed attempts to demolish the bridge.

A demolished bridge becomes Impassable for vehicles and Gun teams, however it is only Very Difficult Going for Infantry and Man-packed-gun teams.

Secure the Crossings

Once secured, pioneers quickly cut all of the wires they could find to make sure the enemy couldn't demolish the bridge remotely. After checking it over, the engineers opened the bridge for business as quickly as possible.

The Attacking players may secure a crossing in the same way as they would normally Take an Objective. Once secured, a bridge cannot be demolished for the rest of the game.

Reserves

The Defenders in the River Rush mission use the normal Reserves rule found on page 268 of the rulebook. When rolling for Reserves, each defending player rolls as normal, and each result of 5+ allows that player to choose a platoon from their own force to arrive on the table.

In for a Penny...

This is it! It's all or nothing. The enemy must be stopped and routed. We must be victorious on the battlefield.

No player is required to take Company Morale Checks. Instead, companies fight to the last in this critical battle. However, platoons still take Platoon Morale Checks as normal.

When an Attacking player's Platoon is destroyed, it is placed in Reserve instead of being permanently removed from play.

The Attacking player may voluntarily destroy platoons and return them to Reserve in the first part of the Starting Step.

Rushed Reserves

There isn't a moment to loose. Get you men back into action as quickly as possible!

At the beginning of the Starting Step, each Attacking player rolls one die for each platoon held in Reserve.

- *On a result of 5+, the platoon arrives from Reserve on the table.*
- *Otherwise, the platoon is still rebuilding itself and remains in Reserve.*

An Attacking player may deploy platoons held in Reserves either from their table edge, or from any secured crossing.

If they arrive from a secured crossing, they must be placed anywhere within 12"/30cm of the centre of the crossing, and may not be placed within 16"/40cm of enemy teams (other than Independent teams).

Like usual, platoons arriving from Reserves are considered to have moved in their Shooting Step.

RIVER RUSH

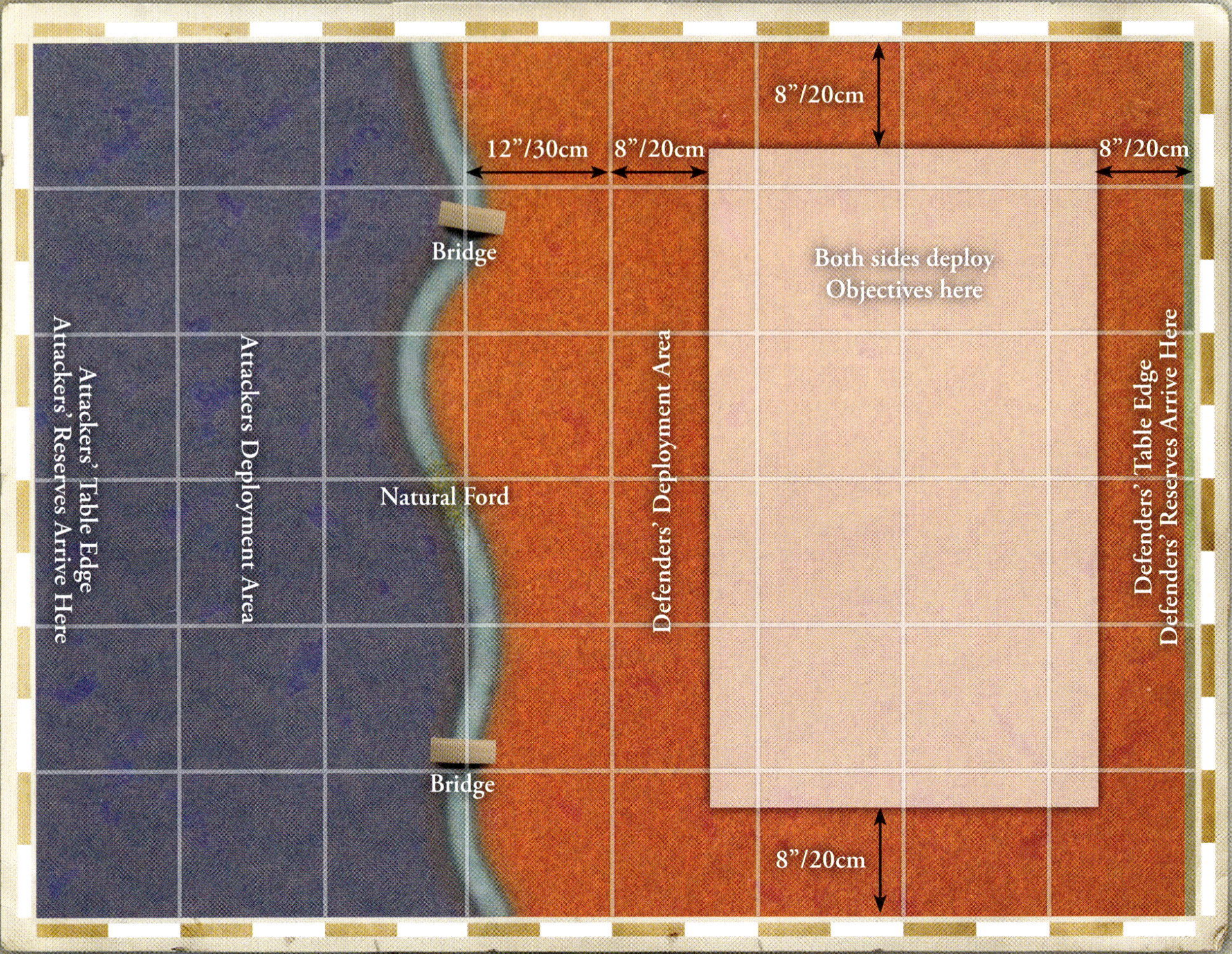
8"/20cm
12"/30cm
8"/20cm
8"/20cm
Bridge
Both sides deploy Objectives here
Attackers' Table Edge
Attackers' Reserves Arrive Here
Attackers Deployment Area
Natural Ford
Defenders' Deployment Area
Defenders' Table Edge
Defenders' Reserves Arrive Here
Bridge
8"/20cm

This mission uses the **Bridge Demolition**, **Secure the Bridges**, **Reserves**, **In For A Penny**, **Rushed Reserves** (see page 107), and the **Prepared Positions** (see page 264 of the rulebook) special rules.

Attackers

We need to take the town to break the enemy's flank wide open. The enemy has secured key bridges that provide the most direct route to the town. There is a fordable pass in the river but we don't want to be bottle-necked into attacking a heavy defended rout. Secure the bridges and take the town as fast as possible.

Defenders

Intelligence has informed us that the enemy is preparing to strike our HQ. If it comes down to the wire you have orders to blow the bridges leading into town, but this is a last resort. Try to fend off the enemy. If you have to blow the bridges leg it back to the city to help defend our HQ from the enemy.

Preparing for Battle

1. Decide on a start and ending time for the game. You will probably need half an hour plus an extra half an hour for every 1000 points on a side. Make sure you have an extra hour available after the end time so that you can finish both players' turns.

2. Set up the table as shown on the map with a river, two bridges, and a natural ford. Then set up a town on the other end. Add forests and hills in the more open areas.

3. The defenders place two objectives in the town. These objectives must be more than 24"/120cm from each other. Then the Attackers place two objectives in the town.

4. Each Defender now nominates at least half of their platoons to be held off the table in Reserve. The remaining platoons are then deployed on the table in the Defenders' Deployment Area.

5. The Attackers now deploy their entire force on the opposite side of the river as the Defenders so that all of their teams are deployed more than 16"/40cm from all enemy teams. Attacking players will use the Rushed Reserves special rule.

6. Both sides, starting with the Defenders, deploy their Independent teams within their own Deployment Areas.

Beginning the Battle

1. The Attackers have the first turn.

2. The Defenders start the game in Prepared Positions.

Ending the Battle

The battle ends as soon as both sides have had the same number of turns and the agreed upon ending time has passed.

If the game has reached a critical moment and both sides want to see what happens next, give each player another turn or two to find out.

Deciding Who Won

If the Attackers hold an Objective when the game ends, they win the battle. The enemy's flank has been split wide open and this victory will lead your armies deep into the enemy's territory.

Otherwise, the Defenders win. The bold effort has been blunted and the enemy's forces severely depleted. The time for counterattack is upon us!